Trichoderma - The Most Widely Used Fungicide

Edited by Mohammad Manjur Shah, Umar Sharif and Tijjani Rufai Buhari

Published in London, United Kingdom

IntechOpen

Supporting open minds since 2005

Trichoderma - The Most Widely Used Fungicide
http://dx.doi.org/10.5772/intechopen.77912
Edited by Mohammad Manjur Shah, Umar Sharif and Tijjani Rufai Buhari

Contributors

Snježana Topolovec-Pintarić, Ramsés González-Estrada, Porfirio Gutierrez-Martinez, Francisco Blancas-Benítez, Cristina Moreno-Hernández, Anelsy Ramos-Guerrero, Angel Fonseca-Cantabrana, Luz Del Carmen Romero-Islas, Jovita Romero-Islas, Rosa Avila-Peña, Beatriz Montaño-Leyva, Syed Ali Nusaibah, Habu Musa, Fernando Cezar Cezar Juliatti, Tâmara Prado De Morais, Anakely A. Rezende, Breno Cezar Marinho Juliatti, Vipin Parkash, Akshita Gaur, Rahul Agnihotri, Ashok Aggarwal, Mohammad Manjur Shah, Hamisu Afiya

First published in London, United Kingdom, 2019 by IntechOpen
IntechOpen is the global imprint of INTECHOPEN LIMITED, registered in England and Wales, registration number: 11086078, The Shard, 25th floor, 32 London Bridge Street
London, SE19SG - United Kingdom
Printed in Croatia

British Library Cataloguing-in-Publication Data
A catalogue record for this book is available from the British Library

Additional hard and PDF copies can be obtained from orders@intechopen.com

Trichoderma - The Most Widely Used Fungicide
Edited by Mohammad Manjur Shah, Umar Sharif andTijjani Rufai Buhari
p. cm.
Print ISBN 978-1-78923-917-1
Online ISBN 978-1-78923-918-8
eBook (PDF) ISBN 978-1-83881-855-5

For EU product safety concerns:
IN TECH d.o.o., Prolaz Marije Krucifikse Kozulić 3, 51000 Rijeka, Croatia, info@intechopen.com or visit our website at intechopen.com.

Meet the editors

Dr. Mohammad Manjur Shah obtained his PhD in Zoology in 2003 from Aligarh Muslim University. He has presented his research findings at several conferences and published articles in various international journals. He completed two post-doctoral fellowships under the Ministry of Science and Technology, Government of India, before becoming a Senior Assistant Professor at Northwest University, Kano, Nigeria, in 2015. Apart from the present book, he has edited four other volumes published by IntechOpen. He is also a reviewer for several journals of international repute.

Dr. Umar Sharif obtained an MSc and PhD from Byero University, Kano, Nigeria. Currently, he is head of the Department of Biological Sciences at Yusuf Maitama Sule University, Kano, Nigeria. Since 1990, Dr. Sharif has been actively involved in teaching as well as various research activities. He has good expertise in the fields of pathology and biology, and is a member of various scientific societies in Nigeria. He has presented his research findings at various conferences and has published papers in journals of international repute.

Dr. Tijjani Rufai Buhari obtained a PhD from the Universiti Putra Malaysia in 2012. At present, he is Deputy Dean of the Faculty of Science at the same university. Since 2007, he has been engaged in teaching and research. His areas of interest include developmental biology, embryology, fisheries and aquaculture, marine biology, environmental microbiology, genetics, general biology, food and nutrition, and food sanitation. In addition to publishing many papers in journals of international repute, Dr. Buhari has attended and presented at several conferences, seminars and workshops.

Contents

Preface

Trichoderma is a genus of fungi present in all soils where they are the most prevalent culturable fungi. They are the most successful biofungicides used in today's agriculture. *Trichoderma* species are non-pathogenic fungi, often found in soil as well as in association with plants. These green-colored fungi are well known for their antifungal and plant-growth-stimulating effects. The species of *Trichoderma* spp. is the most widely used fungus in many areas, including medical, agricultural and industrial applications. This book provides comprehensive information on *Trichoderma* with chapters covering such topics as: taxonomic status, biodiversity, antagonistic properties, application as a bio-control agent, mechanism of bio-control, molecular basis of bio-control, biochemical analysis, effective and up-to-date culture methods, screening of various strains, genetic engineering for enhanced effectiveness in the bio-control application, and the interaction between fungus, plant, and animal.

The book was edited by collaborating expert opinion in the field from different authors from different countries. It consists of six chapters organized into two sections. Section I focuses mainly on identification of *Trichoderma* species, and Section II is concerned with *Trichoderma* as a biological control agent. The book is well illustrated for easy understanding of the important information being discussed.

The editors appreciate all the staff at IntechOpen for making this book project successful. We hope readers will find this volume helpful and useful.

Dr. Mohammad Manjur Shah
Yusuf Maitama Sule University,
Kano, Nigeria

Dr. Umar Sharif and Dr. Tijjani Rufai Buhari
Yusuf Maitama Sule University,
Kano, Nigeria

Section 1

Identification of Trichoderma Species

Chapter 1

Introductory Chapter: Identification and Isolation of *Trichoderma* spp. - Their Significance in Agriculture, Human Health, Industrial and Environmental Application

Mohammad Manjur Shah and Hamisu Afiya

1. Introduction

The genus *Trichoderma* is a diverse group of free-living fungi in the family *Hypocreaceae*, commonly present in all soils [1–6]. These ascomycetes fungi are opportunistic, avirulent plant symbionts inhabiting root ecosystems [3, 7] and parasites on other groups of fungi [2]. They reproduce by chlamydospores and ascospores and proliferate better at mesophilic temperatures (25–35°C) and wide range of pH. Several findings supported this such as [8] who observed no visible growth of conidia at 15°C, but retain growth at 25°C and best results at 30°C [9], evaluated the growth of *Trichoderma* isolates at different temperatures and pH ranges, and reported the highest mycelial growth of *T. hamatum T612*, *T. harzianum T447*, *T. harzianum T969*, and *T. hamatum T614* at 25°C and *T. virens T523* and *Trichoderma* sp. at 30°C. For pH requirement, mycelial growth of *T. hamatum T612*, *T. harzianum T447*, and *T. virens T523* grew best at pH 5 and *T. harzianum T969* and *Trichoderma* sp. at pH 7, while *T. hamatum T614* has best mycelial growth at pH 8. A recent study by [10] reported 25–35°C. Similarly, a pH range of 5.5–8.5 was congenial for *T. harzianum* and *T. hamatum*.

Trichoderma colonizes several ecological niches where they play a vital role; they have been earlier recognized as effective biocontrol agents of plant-pathogenic fungi, producers of secondary metabolites of medical importance [3, 11, 12], and agents of bioremediation. Similarly, their ability to degrade lignocellulosic biomass to produce second-generation biofuels and other value-added products has been widely accepted [3, 12].

2. Identification of *Trichoderma* isolates

Conventional methods for identification of *Trichoderma* spp. using morphological and cultural approach have earlier been used. These include arrangement of conidiophores, phialides, and conidia, while cultural features include linear growth, colony color, growth pattern, and pigmentation of hyphae. The fungus has revealed different morphologies on various cultivation media due to genetic factors and environmental and nutritional factors. Green colony pigmentation after incubation for 7 days at 28°C on potato dextrose agar (PDA) was observed in *Trichoderma* cultures isolated from soil samples in Kliran [5]. Rhizospheric isolates revealed pale or yellowish color of reverse colonies at 25 and 30°C with rapid growth, loosely arranged conidia, and effused conidiation [13]. An ellipsoidal, obovoid, and bowling pin phialides were observed in *Trichoderma* spp. [14]; 10 isolates from groundnut rhizosphere revealed different morphological and microscopic features as shown in **Table 1**.

S/N	Isolate	Colony color	Colony reverse color	Conidiophore character	Phialide character	Conidia shape	Chlamydospore formation
1	GRT-1	Dark green	Amber	Long infrequently branching and verticillate	Frequently paired, lageniform, and divergent	Globose to ellipsoidal	Infrequent, terminal, and intercalary
2	GRT-2	Dull green to bluish green	Colorless	Broad, verticillate, and frequently branching	Lageniform, divergent, terminal philaid more elongated	Sub cylindrical to narrow ellipsoidal	Frequent, intercalary, and terminal
3	GRT-3	White	White	—	—	No conidia	Abundant, terminal, and intercalary
4	GRT-4	Scattered in minute tufts and pale yellow green	Pale yellowish	Rarely branched and verticillate	Cylindrical or slightly inflated and divergent	Ellipsoidal	Frequently intercalary and terminal
5	GRT-5	Dell green to bluish green	Pale yellowish	Broad frequently branching and verticillate	Ampulliform and divergent	Sub cylindrical	Infrequent, intercalary, and terminal
6	GRT-6	Dark bluish green	Uncolored	Infrequently branching and verticillate	Lageniform and convergent	Globose to ellipsoidal	Frequently intercalary and terminal
7	GRT-7	Dark green producing tufts or pustules fringed by sterile mycelium	Dull yellowish	Frequently branching and verticillate	Ampulliform and convergent	Sub globose to obovoid	Infrequent, internally and terminally
8	GRT-8	Dull green to bluish green	Pale yellowish	Narrow verticillate and frequently branching	Ampulliform and divergent	Sub cylindrical to narrow ellipsoidal	Infrequent, intercalary, and terminal
9	GRT-9	Dark bluish green	Uncolored	Infrequently branching and verticillate	Lageniform and convergent	Globose to ellipsoidal	Infrequent, intercalary, and terminally
10	GRT-10	Compute dull, green tufts or pustules	Discolored	Frequently branching and pyramidal structure	Lageniform and divergent	Obovoid	Frequently, intercalary, and terminal

Source: [15].

Table 1.
Morphological characteristics of some Trichoderma isolates.

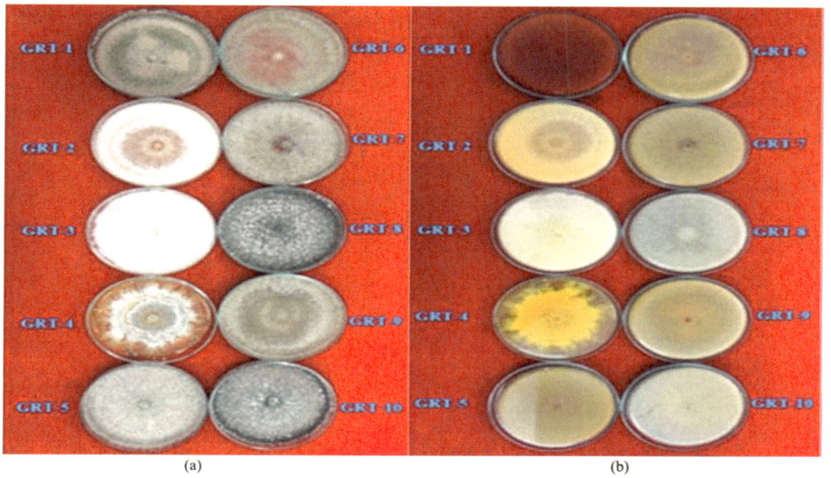

Figure 1.
Trichoderma isolates after 5 days incubation period [15]. (a) Colony color. (b) Colony reverse color.

Isolates GRT-1, GRT-6, and GRT-9 were confirmed as *Trichoderma viride*; GRT-2, GRT-5, and GRT-8 as *Trichoderma koningii*; *GRT-4* as *Trichoderma reesei*; GRT-7 as *Trichoderma harizanum*; and GRT-10 as *Trichoderma aureoviride*, while GRT-3 could not be identified up to species level (**Figure 1**) [15]. However, conventional methods for the identification of *Trichoderma* spp. using morphological and cultural methods are prone to error and poor documentation of isolates at the culture collection since isolates are not "adequately differentiated" [16]. The use of molecular phylogenetic features coupled with the conventional methods is a better approach for verification of isolates and identification of novel strains [8] than morphological features alone, as anamorph and teleomorph used for defining species have reached their limits [17]. Several literatures have reported various techniques of molecular characterization to confirm *Trichoderma* isolates, such as "amplifying and analyzing the sequences of internal transcribed spacer gene (ITS) 1 and 2 and translation elongation factor 1-alpha (tef1) encoding gene" and BLAST interface in TrichOKEY and TrichoBLAST [16]; "restriction fragment length polymorphism (RFLP) and DNA sequencing" [18–20]; random amplified polymorphic DNAs (RAPD) and rDNA sequencing [14, 21]; sequence-related amplified polymorphism (SRAP) marker [7]; etc.

3. Isolation media for *Trichoderma*

The growth of *Trichoderma* has been screened on different culture media for various studies using available, relatively cheaper supporting media such as corn meal agar, oat meal agar, potato dextrose agar, Czapek's Dox agar, special nutrient media, carrot agar, rose Bengal agar, selective media, etc. However, selective media favor growth of *Trichoderma* strains over other fungi and hence preferred for easy identification of *Trichoderma* isolates over rapidly growing fungi that may overlap it [22].

3.1 *Trichoderma* selective media (TSM)

Trichoderma selective medium (TSM) is recognized for quantitative isolation of *Trichoderma* spp. from soil. It is composed of low glucose level for rapid growth and sporulation of the fungus. Chloramphenicol is used to inhibit the growth of bacteria, while pentachloronitrobenzene, p-dimethylaminobenzenediazo sodium sulfonate, and rose bengal are used as selective fungal inhibitors [22].

3.1.1 TSM recipe

1. 0.2 g of $MgSO_4{\cdot}7H_2O$.
2. 0.9 g of K_2HPO_4.
3. 0.15 g of KCl.
4. 1.0 g of NH_4NO_3.
5. 3.0 g of glucose.
6. 0.15 g of rose bengal.
7. 20 g of agar.
8. 0.25 g of chloramphenicol.
9. 0.3 g of p-dimethylaminobenzenediazo sodium sulfonate.
10. 0.2 g of pentachloronitrobenzene.

Recipe is dissolved in 1000 ml distilled water and autoclaved at 121°C, 1.4 kg cm^{-1} for 15 min. Then add 0.25 g chloramphenicol and 0.2 g pentachloronitrobenzene into the solution. Keep/store media at 45°C to prevent solidification.

3.2 *Trichoderma harzianum* selective medium (THSM)

Selection of THSM enables comparison between aggressive and non-aggressive *Trichoderma* groups. The antimicrobials chloramphenicol, streptomycin, quintozene, and propamocarb are added to the medium to highly select *T. harzianum* in compact colonies without visible contamination [23].

3.2.1 Recipe for THSM

1. 0.2 g of $MgSO_4{\bullet}7H_2O$.
2. 0.9 g of K_2HPO_4.
3. 1.0 g of NH_4NO_3.
4. 0.15 g of KCl.
5. 0.15 g of rose bengal.
6. 3 g of glucose.
7. 20 g of agar.
8. 950 mL of distilled water.

Media is autoclaved at 121°C, 1.4 kg cm^{-1}, for 15 min, and 0.25 g of chloramphenicol, 9.0 mL of streptomycin, 1.2 mL of propamocarb, and 0.2 g of quintozene are added.

3.3 Rose Bengal agar (RBA)

RBA is a nonselective medium for isolation of *Trichoderma* which is developed by Jarvis in 1973, for enumeration of molds and yeasts from food. The medium is suitable with protein foods and tolerates high temperatures. Chloramphenicol or chlortetracycline is added to suppress the growth of bacteria [24].

3.3.1 Recipe for rose Bengal agar (RBA)

1. Mycological peptone 5.0 g
2. Rose bengal 0.05 g
3. Glucose 10.0 g
4. Chloramphenicol 0.1 g
5. Dipotassium phosphate 1.0 g
6. Agar 15.0 g
7. $MgSO_4.7\ H_20$ 0.5 g
8. Distilled 1000 mL

Add the ingredients to the distilled water and boil to dissolve completely. Add 10 mL of chloramphenicol or chlortetracycline; shake and autoclave at 121°C for 15 min. Store in the dark at 4°C for further use.

4. Method for isolation of *Trichoderma*

Several methods are available for the isolation of *Trichoderma*; however, one of the commonest methods reported in literature is the serial dilution of samples [22, 25–28]. This technique is simple, cost-effective, and appropriate to handle large samples.

Soil samples are collected, air dried, and ground into powder. Stock solution of sample is prepared by dissolving 10 g of powdered soil sample into 90 mL of distilled water. Next, serial dilution of samples were prepared as 10^{-1}, 10^{-2}...10^{-5}. One milliliter of each of the prepared dilution is spread evenly on a suitable medium on a petri dish at 28 ± 1°C for 7 days.

5. Agricultural significance of *Trichoderma* spp.

Continuous use of chemical pesticides to manage fungal pathogens (which are known to cause major diseases in agriculture) has led to destruction of soil structure, soil infertility, and accumulation of toxic compounds on crops. Moreover, "chemical fungicides have less influence on pathogens due to their diversity, adaptability and increasing resistance" [4]. Various microbial biocontrol agents serve a solution for management of the aforementioned to attain a sustainable agriculture for future generations [29].

Knowledge about biocontrol potential of the fungus *Trichoderma* spp. has been recognized as early as 1920 [30, 31], although it received researcher's interest with advances in genetic engineering [12]. This technology has made it easy to isolate, characterize, clone, sequence, and express the roles of specific genes in the biocontrol mechanism. The genes encoding the enzymes play vital roles in biotic and abiotic stress tolerance, growth of hyphae, degradation of cell wall, and antagonistic activity against plant pathogens [29]. *Trichoderma harzianum* (Th. Azad) and *Trichoderma viride* (01PP) are used as biopesticides and biofertilizers [32, 33], growth promoters, and inducers of disease resistance in plants [12, 33]. The former is the main antagonist utilized in management of plant diseases in agriculture [34, 35] due to its cost-effectiveness and minimal effects on the ecological balance [34].

Trichoderma is efficient in improving vegetative growth of plants and nutrient content of soil through decomposition and biodegradation [33]. Active substance such as fungal spores is applied as foliar sprays and pre- and post-planting treatments, during watering and transplanting. *Trichoderma*-based products are marketed worldwide and applied in fields, nurseries, and horticulture for management of fungal soil-borne pathogens such as *Pythium* and *Rhizoctonia* [33, 35]. It is a safe and environmentally friendly method to reduce the detrimental effects of chemical pesticides [36].

Various articles reported on the role of *Trichoderma* spp. as antagonist to plant pathogens such as *T. harzianum*, *T. asperellum*, and *T. virens* against *Phytophthora capsici* in red pepper [16]; *Trichoderma* isolates against *Sclerotium rolfsii*, *Colletotrichum gloeosporioides*, *C. capsici* [37], *S. minor* and *S. sclerotiorum* in the in vitro experiments [38]; *T. atroviride SY3A* and *T. harzianum SYN* were effective biological control agents of *R. solani* damping-off of cucumber [39]; *Trichoderma* isolates were antagonist to soil-borne phytopathogenic fungi (*Fusarium graminearum*, *Rhizoctonia solani*, *Macrophomina phaseoli*, and *Phytophtora cactorum*) [9]; *Trichoderma* species was antagonist to anthracnose of strawberry [5]; *Trichoderma* isolates inhibit and control the growth of *Fusarium oxysporum* with *Trichoderma harzianum* being the most effective [40]; *T. viride*, *T. polysporum*, and *T. harzianum* inhibit more than 60% growth of *C. paradoxa* [19]; *T. hamatum LU593* and *T. virens LU556* delayed aphids manifestation on cabbage [41]; *Trichoderma* isolates against *Sclerotium rolfsii* [6]; *Trichoderma* isolates against *Fusarium sambucinum* [42]; and *Trichoderma* spp. and *Bacillus* spp. in seed treatment against root knot nematode *Meloidogyne javanica* [43].

6. Mechanism of biological control

Several researches have revealed the mechanism of biocontrol in *Trichoderma* via mycoparasitism (by coiling around the host, formation of appressoria and breakdown of the host cell wall), antibiosis, and competition for resources (space and nutrients). A peptaibol from *Trichoderma* may induce apoptosis in plant-pathogenic fungi through complex mechanisms. Trichokonins, a type of peptaibol from *Trichoderma pseudokoningii SMF2*, produced a molecular biocontrol mechanism which efficiently induces apoptosis in fungal cells. Apoptotic hallmarks such as the deposition of cytoplasmic vacuoles, presence of reactive oxygen species, breakage of DNA molecule, and exposure of phosphatidylserine were observed in *Fusarium oxysporum* cells treated with Trichokonins [44]. Mycoparasitism has been reported in *Trichoderma* species antagonist to Anthracnose disease of strawberry.

Microscopic examination revealed that the hyphae of *Trichoderma* "grew alongside and coiled compactly around the hyphae of the fungal pathogen isolates" [5].

Recent development has further shown the significance of *Trichoderma* to induce systemic/localized resistance in plant by colonizing the root epidermis and subsequent release of bioactive metabolites, to transform the transcriptome and proteome of resultant plants [11, 30]. It further revealed that metabolites/enzymes produced in *Trichoderma* such as chitinase and β-1,3 glucanase as recorded in *T. viride* and *T. harzianum*, respectively [37]; cellulases and hemicellulases in *Trichoderma* spp. [12, 45]; and proteases and lipases [46] are responsible for breaking down the component of the fungal cell wall to reduce the integrity of the pathogen cell. Relevant biocontrol processes such as production of hydrolytic enzymes and antifungal metabolites and the formation of infection structures are controlled by heterotrimeric G proteins and mitogen-activated protein (MAP) kinases. Similarly, induction of plant systemic resistance in *Trichoderma virens* and hyperosmotic stress response in *Trichoderma harzianum* are related to MAPK signaling [47], while biocontrol associated with coiling and chitinase production in *Trichoderma* is regulated by internal cAMP level.

7. *Trichoderma* in human health

"*Trichoderma* species are possible source of important antimicrobial agents against gram negative, positive, fungi and yeast" [48]. Earlier in 1995, isolated peptides from *Trichoderma* strains showed antibacterial activity against *S. aureus* [49]; *T. harzianum* produced 44.06 μg/mL of the well-known antifungal drug, cyclosporine [50]. Similarly, 'Trichodermanins C–E (1–3), new diterpenes with a rare fused 6-5-6-6ring system, have been isolated from a fungus *Trichoderma harzianum*" detached from a marine sponge *H. okadai*. Cytotoxicity assay using three cancer cell lines showed significant activity in 1 [51]. Broth extracts of *Trichoderma* species (*Trichoderma harzianum*, *Trichoderma longibrachiatum*, and *T. koningii*) showed antifungal and antibacterial activity against *Paecilomyces variotii*, *Penicillium notatum*, *Nematospora coryli*, *Mucor miehei*, *Bacillus brevis*, *Bacillus subtilis*, *Enterobacter dissolvens*, and *Sarcina lutea* using agar disk diffusion method [48].

8. Industrial and environmental applications of *Trichoderma* spp.

Trichoderma are utilized in the production of low-cost enzymes for applications in food, pulp, and paper and textile industries to generate biofuel. The biotechnological workhorse of *Trichoderma* is the production of cellulases [52], in addition to the extracellular laccase [53]. Together with cellulases, endoglucanases (EG1, EG2, EG3, and EG5), and β-glucosidase, these enzymes catalyze the breakdown of ligninolytic biomass to produce an important industrial enzyme for the production of second-generation biofuels and other value-added products such as fermentable sugars, organic acids, solvents, drink softeners, etc. [54]. The production of proteins such as heterogenous proteins from cellobiohydrolase I (cbhI), a strong and inducible promoter of the gene encoding the major cellulose [55], and hydrophobins HFBI and HFBII [56–58] for industrial applications has been reported. Research efforts have focused on increasing enzyme yield and other valuable products from *Trichoderma* spp. through genome sequencing [45]. The highest yield of 1.4 g hydrophobins HFBI and 0.24 g hydrophobins HFBII per liter was obtained from the fungus through genetic manipulation on glucose-containing culture medium [59]; up to 40 mg/l of chymosin was produced from a transformed strain [55]; newly constructed *Trichoderma* strains designed for specific industrial applications such as biofinishing and biostoning of cotton increased the cellulose CBHI by 1.5-fold and CBHII by fourfold as to the main strain. These were further increased to 1.6-fold CBHI and 3.4-fold CBHII by transformation as compared to the host strain. In addition, CBHII proteins were produced by the gene promoter (*Trichoderma* cellulases).

9. Conclusion

Trichoderma spp. is one of the frequently isolated fungal genera from soil and plant roots that have been extensively studied for their vast metabolites with various applications (agricultural, industrial, health, etc.).

In the field of agriculture, *Trichoderma* are suitable antimicrobial agents against pathogenic bacteria, fungi, and yeast. Similarly, they play a vital role in improving the vegetative growth of plants and nutrient content of soil through decomposition and biodegradation. It is a safe, cost-effective, and environmentally benign technology to attain a sustainable agriculture.

In the field of medicine, different metabolites of medical importance have been reported from *Trichoderma*. Earlier in 1995, isolated peptides from *Trichoderma* strains showed antibacterial activity against *S. aureus* [49]. *T. harzianum* produced 44.06 μg/mL of the well-known antifungal drug, cyclosporine [50].

Cellulases, an important industrial enzyme from *Trichoderma*, are essential in the breakdown of biomass to produce second-generation biofuels and other value-added products such as fermentable sugars, organic acids, solvents, drink softeners, [54] etc., in addition to the laccase production for textile industries. With advances in genetic engineering, efforts are focused on designing new strains of *Trichoderma* spp. through genome sequencing for production of novel metabolites of various applications.

Author details

Mohammad Manjur Shah* and Hamisu Afiya
Department of Biological Sciences, Yusuf Maitama Sule University, Kano, Nigeria

*Address all correspondence to: mmanjurshah@gmail.com

References

[1] Samuels GJ. *Trichoderma*: A review of biology and systematics of the genus. Mycological Research. 1996;**100**(8):923-935. DOI: 10.1016/S0953-7562(96)80043-8

[2] Harman GE, Howell CR, Viterbo A, Chet I, Lorito M. *Trichoderma* spp.—Opportunistic, avirulent plant symbionts. Nature Reviews. Microbiology. 2004;**2**(1):43-56

[3] Schuster A, Schmoll M. Computer-assisted corneal topography. High-resolution graphic presentation and analysis of keratoscopy. Applied Microbiology and Biotechnology. 2010;**87**:787-799

[4] Matei S, Matei G-M, Cornea P, Popa G. Characterization of soil *Trichoderma* isolates for potential biocontrol of plant pathogens. Factori şi procese pedogenetice din zona temperată. 2011;**10**:29-37

[5] Mutia D, Prilya F. Exploration of *Trichoderma* spp. and fungal pathogen that causes a strawberry anthracnose and examination of in vitro antagonistic activity. Biotika. 2017;**5**(18):58-68

[6] Mohan PN. Studies on antagonistic potential of *Trichoderma* spp. from saline soils [Internet]. Agricultural University; 2017. Available from: http://krishikosh.egranth.ac.in/handle/1/5810030745

[7] Digamber PS. Geographical diversity analysis of *Trichoderma* spp. isolates based on sequence related amplified polymorphism (SRAP) marker. Latur: College of Agricultural Biotechnology; 2017

[8] Soesanto L, Sri-Utami D, Rahayuniati RF. Morphological characteristics of four *Trichoderma* isolates and two endophytic *Fusarium* isolates. Canadian Journal on Scientific and Industrial Research. 2011;**2**(8):294-306

[9] Hajieghrari B, Torabi-giglou M, Mohammadi MR, Davari M. Biological potential of some Iranian *Trichoderma* isolates in the control of soil borne plant pathogenic fungi. African Journal of Biotechnology. 2008;7(8):967-972

[10] Shravan kashid V. Studies on Compatibility of *Trichoderma* spp. with Agrochemicals. Parbhani: Vasantrao Naik Marathwada Krishi Vidyapeeth; 2017

[11] Mukherjee PK, Horwitz BA, Herrera-Estrella A, Schmoll M, Kenerley CM. *Trichoderma* research in the genome era. Annual Review of Phytopathology. 2013;**51**(1):105-129. Available from: http://www.annualreviews.org/doi/10.1146/annurev-phyto-082712-102353

[12] Błaszczyk L, Siwulski M, Sobieralski K, Lisiecka J, Jędryczka M. *Trichoderma* spp.—Application and prospects for use in organic farming and industry. Journal of Plant Protection Research. 2014;**54**(4):309-317

[13] Sharma KK, Singh US. Cultural and morphological characterization of rhizospheric isolates of fungal antagonist *Trichoderma*. Journal of Applied and Natural Science. 2014;**6**(2):451-456

[14] Choi I-Y, Hong S-B, Yadav MC. Molecular and morphological characterization of green mold, *Trichoderma* spp. isolated from oyster mushrooms. Mycobiology. 2003;**31**(2):74-80. Available from: https://www.mycobiology.or.kr/Synapse/Data/PDFData/0184MB/mb-31-74.pdf

[15] Sekhar YC, Ahammed SK, Prasad TNVKV. Identification of *Trichoderma*

species based on morphological characters isolated from rhizosphere of groundnut (*Arachis hypogaea* L). International Journal of Science, Environment and Technology. 2017;**6**(3):2056-2063

[16] Savitha MJ, Sriram S. Morphological and molecular identification of *Trichoderma* isolates with biocontrol potential against *Phytophthora blight* in red pepper. Pest Management In Horticultural Ecosystems. 2015;**21**(2):194-202

[17] Samuels GJ. *Trichoderma*: Systematics, the sexual state, and ecology. The Nature and Application of Biocontrol Microbes II: Trichoderma spp. 2006;**96**(2):1-12

[18] Bourguignon E. Ecology and diversity of indigenous *Trichoderma* species in vegetable cropping systems. Lincoln University; 2008

[19] Kannangara S, Dharmarathna R, Jayarathna DL. Isolation, identification and characterization of *Trichoderma* species as a potential biocontrol agent against *Ceratocystis paradoxa*. The Journal of Agricultural Science. 2016;**12**(1):51-62. Available from: http://jas.sljol.info/article/10.4038/jas.v12i1.8206/

[20] Hatvani L. Mushroom pathogenic *Trichoderma* species: Occurrence, biodiversity, diagnosis and extracellular enzyme production. University of Szeged; 2008

[21] Narayan K, Kotasthane A. Genetic relatedness among *Trichoderma* isolates inhibiting a pathogenic fungi *Rhizoctonia solani*. Journal of Biotechnology. 2006;**5**:580-584

[22] Elad Y, Chet I, Henis Y. A selective medium for improving quantitative isolation of *Trichoderma* spp. from soil. Phytoparasitica. 1981;**9**(1):59-67

[23] Williams J, Clarkson JM, Mills PR, Cooper RM. A selective medium for quantitative reisolation of *Trichoderma harzianum* from *Agaricus bisporus* compost. Applied and Environmental Microbiology. 2003;**69**(7):4190-4191

[24] Rose bengal chloramphenicol (RBC) agar. International Journal of Food Microbiology. Elsevier; 1987;**5**(3):261-262

[25] Mustafa A, Khan AM, Inam-ul-Haq M, Pervez AM, Umar U-D. Usefulness of different culture media for in-vitro evaluation of *Trichoderma* spp. against seed-borne fungi of economic importance. Pakistan Journal of Phytopathology. 2009;**21**(1):83-88

[26] Khandelwal M, Datta S, Mehta J, Naruka R, Makhijani K, Sharma G, et al. Isolation, characterization & biomass production of *Trichoderma viride* using various agro products—A biocontrol agent. Advances in Applied Science Research. 2012;**3**(6):3950-3955

[27] Roy MK, Hembram S, Debnath A. Effect of different media and pH on growth and sporulation of different native *Trichoderma* spp. An International Quarterly Journal of Life Sciences. 2015;**10**(4):1833-1837

[28] Iqbal S, Ashfaq M, Malik AH, Khan KS, Mathew P. Isolation, preservation and revival of *Trichoderma viride* in culture media. Journal of Entomology and Zoology Studies. 2017;**5**(3):1640-1646

[29] Sharma P, Vignesh KP, Ramesh R, Saravanan K, Deep S, Sharma M, et al. Biocontrol genes from *Trichoderma* spp.: A review. African Journal of Biotechnology. 2011;**10**(86):19898-19907

[30] Harman G. Overview of mechanisms and uses of *Trichoderma* spp. In: The Nature and Application of Biocontrol Microbes II: *Trichoderma* spp. Overview. Phytopathology. 2006;**96**(2):190-194

[31] Harman GE. Overview of mechanisms and uses of *Trichoderma* spp. Phytopathology. 2006;**96**(2): 190-194. Available from: http://apsjournals.apsnet.org/doi/10.1094/PHYTO-96-0190

[32] Kumar V, Shahid M, Srivastava M, Sonika P, Singh A, Sharma A. Role of secondary metabolites produced by commercial *Trichoderma* spp. and their effect against soil borne pathogens. Biosensors. 2014;**03**(01):1-5. Available from: http://omicsonline.com/open-access/role-of-secondary-metabolites-produced-by-commercial-trichoderma-species-and-their-effect-against-soil-borne-pathogens-2090-4967-3-108.php?aid=30417

[33] Woo SL, Ruocco M, Vinale F, Nigro M, Marra R, Lombardi N, et al. *Trichoderma*-based products and their widespread use in agriculture. The Open Mycology Journal. 2014;**8**(1):71-126. Available from: http://benthamopen.com/ABSTRACT/TOMYCJ-8-71

[34] Monte EL. *Trichoderma* in organic agriculture. In: Proceedings V World Avocado Congress. 2003. pp. 725-733

[35] Kumar S, Thakur M, Rani A. *Trichoderma*: Mass production, formulation, quality control, delivery and its scope in commercialization in India for the management of plant diseases. African Journal of Agricultural Research. 2014;**9**(53):3838-3852. Available from: http://www.academicjournals.org/AJAR

[36] Hermosa R, Botella L, Montero-Barrientos M, Alonso-Ramírez A, Arbona V, Gómez-Cadenas A, et al. Biotechnological applications of the gene transfer from the beneficial fungus *Trichoderma harzianum* to plants. Plant Signaling & Behavior. 2011;**6**(8):1235-1236

[37] Kumar K, Amaresan N, Bhagat S, Madhuri K, Srivastava RC. Isolation and characterization of *Trichoderma* spp. for antagonistic activity against root rot and foliar pathogens. Indian Journal of Microbiology. 2012;**52**(2):137-144

[38] Elias LM, Domingues MVPF, De Moura KE, Harakava R, Patricio RA. Selection of *Trichoderma* isolates for biological control of *Sclerotinia minor* and *S. sclerotiorum* in lettuce. Summa Phytopathologica. 2016;**42**(23):216-221

[39] Yobo KS. Biological control and plant growth promotion by selected *Trichoderma* and *Bacillus* species. KwaZulu-Natal University; 2005

[40] Shankar GK. Induction of Systemic Resistance in Tomato by using *Trichoderma*. India: PhD Thesis submitted to PhD Thesis submitted to Mahatma Phule Krishi Vidyapeeth; 2015. 219 pp. Available from: http://krishikosh.egranth.ac.in/bitstream/1/95336/1/GHUTUDKAR.pdf

[41] Pus W. Plant-Mediated Effects of *Trichoderma* spp. and *Beauveria bassiana* Isolates on Insect and Pathogen Resistance. New Zealand: Lincoln University; 2017

[42] Ru Z, Di W. *Trichoderma* spp. from rhizosphere soil and their antagonism against *Fusarium sambucinum*. African Journal of Biotechnology. 2012;**11**(18):4180-4186. Available from: http://www.academicjournals.org/AJb/abstracts/abs2012/1Mar/Ru and Di.htm

[43] Chinheya CC. Use of *Trichoderma* and *Bacillus* isolates as seed treatments against the root knot nematode *Meloidogyne Javanica* (Chitwood). University of KwaZulu-Natal; 2015

[44] Shi M, Chen L, Wang X, Zhang T, Zhao P, Song X, et al. Antimicrobial peptaibols from *Trichoderma pseudokoningii* induce programmed cell death in plant fungal pathogens. Microbiology. 2012;**158**:166-175

[45] Seidl V, Seiboth B. *Trichoderma reesei*: Genetic approaches to improving strain efficiency. Biofuels. 2010;**1**(2):343-354

[46] Jorge L. *Trichoderma* strains as biocontrol agents. Advancements in Genetic Engineering. 2014;**03**(01):4172. Available from: http://www.omicsgroup.org/journals/trichoderma-strains-as-biocontrol-agents-2169-0111.1000e110.php?aid=25241

[47] Zeilinger S, Omann M. *Trichoderma* biocontrol: Signal transduction pathways involved in host sensing and mycoparasitism. Gene Regulation and Systems Biology. 2007;**1**:227-234

[48] Tarus PK, Chhabra SC, Langat-Thoruwa C, Wanyonyi AW. Fermentation and antimicrobial activities of extracts from different spp. of *Trichoderma*. African Journal of Health Sciences. 2004;**11**:33-42

[49] Goulard C, Hlimi S, Rebuffat S, Bodo B. Trichorzins HA and MA, antibiotic peptides from *Trichoderma harzianum* I. Fermentation, isolation and biological properties. Journal of Antibiotics (Tokyo). 1995;**48**(11):1248-1253

[50] Azam A, Anjum T, Irun W. *Trichoderma harzianum*: A new fungal source for the production of cyclosporin A. Bangladesh Journal of Pharmacology. 2012;**7**:33-35

[51] Yamada T, Suzue M, Arai T, Kikuchi T, Tanaka R. Trichodermanins C–E, new diterpenes with a fused 6-5-6-6 ring system produced by a marine sponge-derived fungus. Marine Drugs. 2017;**15**:169

[52] Schuster A, Schmoll M. Biology and biotechnology of *Trichoderma*. Applied Microbiology and Biotechnology. 2010;**87**:787-799

[53] Cázares-García S, Vázquez-Garcidueñas M, Vázquez-Marrufo G. Structural and phylogenetic analysis of laccases from *Trichoderma*: A bioinformatic approach. PLoS One. 2013;**8**(1):e55295

[54] Kumar R, Singh S, Singh O. Bioconversion of lignocellulosic biomass: Biochemical and molecular perspectives. Journal of Industrial Microbiology & Biotechnology. 2008;**35**(5):377-391

[55] Uusitalo AH, Penttila BM, Knowles J, Novel Fungal A. Expression system: Secretion of active calf chymosin from the filamentous fungus *Trichoderma reesei*. Biotechnology. 1989;**7**:596-603

[56] Linder M, Selber K, Nakari-Setälä T, Qiao M, Kula MR, Penttilä M. The hydrophobins HFBI and HFBII from *Trichoderma reesei* showing efficient interactions with nonionic surfactants in aqueous two-phase systems. Biomacromolecules. 2001;**2**(2):511-517

[57] Askolin S, Nakari-Setälä T, Tenkanen M. Overproduction, purification, and characterization of the *Trichoderma reesei* hydrophobin HFBI. Applied Microbiology and Biotechnology. 2001;**57**(1-2):124-130

[58] Mustalahti E, Saloheimo M, Joensuu JJ. Intracellular protein production in *Trichoderma reesei (Hypocrea jecorina)* with hydrophobin fusion technology. New Biotechnology. 2013;**30**(2):262-268. DOI: 10.1016/j.nbt.2011.09.006

[59] Askolin S. Characterization of the *Trichoderma reesei* hydrophobins HFBI and HFBII. VTT Publications. Helsinki University of Technology; 2006

Chapter 2

Trichoderma: Invisible Partner for Visible Impact on Agriculture

Snježana Topolovec-Pintarić

Abstract

Species of genus *Trichoderma* may benefit as plant pathogen control agent (mycofungicide) and plant growth promoter (biofertilizer) and their application may lower the production costs and environmental impact. Direct effects of these fungi on plant growth and development are crucially important for agricultural uses and for understanding the roles of *Trichoderma* in natural and managed ecosystems. The *Trichoderma* potential as bioagent is utilized through the commercial production of *Trichoderma*-based product. Commercial products of *Trichoderma*-based biofungicides account for about 60% of the biofungicide market, while the availability and dispersion of *Trichoderma*-based biofertilizers are more widespread than commonly known with a tendency to expand due to the easier registrations. Limiting factors for availability of commercial products are expensiveness of registration requirements as they must be registered as pesticides, especially patenting, efficacy testing, toxicological, and biosafety testing. This chapter intends to give insight into agricultural importance of *Trichoderma* and current status of implementation of *Trichoderma* products in developing and in the developed countries.

Keywords: biocontrol, biotechnological patent, microbial products, pesticide

1. Introduction

The credo of fabulous Spanish architect Antoni Gaudi that *Anything created by human beings is already in the great book of nature* is true for the exploitation of *Trichoderma* benefits. Since the early 1930s, when Weindling reported that *T. lignorum* produces and excretes a "lethal principle" in the surrounding, the scientists become involved in investigation of antifungal ability of various *Trichoderma* species, although *T. harzianum* arisen as the most prominent species of the genus. Today, their agricultural importance is good antagonistic abilities against soil born plant pathogenic fungi, thanks to different mechanisms of antagonism: the production of antifungal metabolites (antibiosis), competition for space and nutrients, induction of defense responses in plant and mycoparasitism. Along with revelation of diverse antifungal mechanisms of *Trichoderma*, the ability to promote plant growth and to increase plant height, leaf area and dry weight were perceived. First, this ability was treated as side effect of suppression of plant pathogenic fungi which leading to stronger root growth and nutrient uptake. Also, positive influence of *Trichoderma* to a faster germination and increase in percentage of emergency were perceived. Nowadays, *Trichoderma* species are considered as opportunistic plant symbionts because they colonize root surface and even penetrate into the epidermis of root tissue and a few cell layers below this level establishing pseudomycorrhizal relationship with plant host. This intimate relationship is what induces localized and systemic resistance plant responses to pathogen attack. For the *Trichoderma*, abundant healthy roots are environment where it grows and proliferates best owing to the main carbohydrates secreted by plant roots. Furthermore, roots are resort of plant pathogenic fungi and nematodes, the target for *Trichoderma* as mycoparasite and nematophagous. The plants also benefit from this relationship through increased root and shoot growth and increased macro- and micronutrient uptake. Therefore, *Trichoderma* may be benefit as growth promotant (biofertilizer) as well as pathogen control agent (mycofungicide), and their application may lower the production costs and environmental impact. Recently, it is recognized that *Trichoderma* positive effect on plant growth is independent ability and equally remarkable and significant as its antifungal ability because growth enhancement has been observed in the absence of any detectable disease and in sterile soil. Therefore, today is considered that the direct effects of these fungi on plant growth and development are crucially important for agricultural uses and for understanding the roles of *Trichoderma* in natural and managed ecosystems.

To exploit *Trichoderma* benefits, it must be isolated from soil, studied, and encapsulated in formulation which will allow application into soil. But, reintroduction to soil, even the most strongly rhizosphere competent such as *Trichoderma* can be difficult. *Trichoderma* reintroduced into soil must compete with spectrum of rhizosphere microbes while trying to colonize available sites along the plant roots. Therefore, it needs to be applied in low cost but high density inocula engineered to maintain fungal propagule viable during the transport, storage, and application. To accomplished mentioned goals and effective dispersal of fungal inocula, it is necessity to choose the fungal inoculum carrier and the type of formulation. The *Trichoderma* potential as bioagent is utilized through the commercial production of *Trichoderma*-based biofungicides, which account for about 60% of the biofungicide market. The availability and dispersion of *Trichoderma*-based biofertilizers are more widespread than commonly known with a tendency to expand due to the easier registrations because they are not registered as pesticides. Limiting factors for availability of commercial products are expensiveness of registration requirements, especially patenting, efficacy testing, toxicological, and biosafety testing.

2. Historical and commercial background

In 1794 the mycologist C. H. Persoon proposed and named the genus *Trichoderma* after mycelial appearing, like hairy (Gr. thrix, genitive trikhos) cowering on decaying wood surface (Gr. derma). Association with teleomorphs in *Hypocrea* was done by Tulasne brothers in 1865. The genus came in focus after Weindling's article about *T. lingorum* as parasite of soil fungi, followed by another article in 1934 about parasitism on *Rhizoctonia solani* helped by some kind of a toxic compound [1, 2]. In that article, Weindling gave definition of antibiosis as suppressive mechanism based on production on secondary metabolites with antimicrobial effect and also named this *T. lingorum* lethal metabolite, gliotoxin. Further Weindling's papers defined biocontrol of plant pathogens through the *Trichoderma* strains and their following unique mechanisms: mycoparasitism, competition for area and nutrients in the rhizosphere and antibiosis, and production of antibiotics [3–5]. Existence of volatile organic compounds produced by *Trichoderma* that can inhibit the growth of fungi responsible for wood decay was published by Dennis and Webster in 1970 [6]. Since then, species of genus *Trichoderma* become among the most commonly studied biocontrol microbes and are presently marketed as active ingredients of biopesticides, biofertilizers, plant growth enhancers, and stimulants of natural resistance.

The *Trichoderma* potential as biocontrol agent is utilized through the commercial production of biofungicides and always registers for use as microbial fungicides. *Trichoderma* based biofungicides are today presented with more than 250 products available worldwide and account for about 60% of the international biofungicide market. The leading country in terms of enormous use of *Trichoderma* products is India which comprises 90% of Asian market [7]. Following is Brazil with the greatest production on the South and Central America. For example, in Venezuela and Cuba, the development and use of *Trichoderma*-based products are government supported and officially recommended [8]. Prevalent species in majority of *Trichoderma*-based products is *T. harzianum* (83%), of which 55% of these are combined with *T. viride* and 28% are with *T. koningii* [7]. Widely used *T. harzianum* and *T. viride* are mostly applicate as soil treatment in around 87 various crops against 70 soil-borne and 18 foliar-borne pathogens, mostly fungi [9].

The first *Trichoderma* biofungicide commercialized worldwide was Trichodex produced by Makhteshim Agan Industries (Beersheba, Israel). Trichodex is based on *T. harzianum* isolate T39 studied since 1986 by its creator Dr. Yigal Elad of the Volcani Center, Israel where research and development were carried out [10]. Worked as a biopesticide researcher, Dr. Elad had primarily studying the biological control of economically significant plant pathogens primary gray mold causal, ascomycete fungus *Botrytis cinerea*. Early contact with industry company and sign of agreement with the aim of developing a biocontrol preparation for the control of gray mold were crucial, and Dr. Elad's collaboration with Makhteshim-Agan Industries led to the development and launch of the biopesticide product, Trichodex. Around the world, Makhteshim carried out efficacy trials of Trichodex in controlling of gray mold in vineyards in more than 130 experiments on 34 varieties, under diverse commercial conditions around the world [11]. In 1993, Trichodex has been registered in countries such as Argentina, Australia, Bulgaria, California, Chile, Colombia, Croatia, Cyprus, Greece, Guatemala, Hungary, Israel, Italy, Morocco, Paraguay, Romania, Turkey, Slovenia, South Africa, and the USA. Trichodex was in Croatia registered as contact antibiotic fungicide with enzymatic activity against *B. cinerea* on grapevine and strawberries. In Croatia, microplots with Trichodex were carried out until 1999 in vineyards of famous winegrowing region Kutjevo, situated in the continental part of Croatia were the gray mold disease inflicts damages of 50–60%. Trichodex shown efficacy in interval of 13–55% depending on weather conditions in year which was satisfactory control and occasionally equally efficient as synthetic fungicide Kidan (a.i. iprodione, Bayer, Germany) registered in Croatia at that time [12, 13]. In 1990s, beside Volcani center (Israel) Research institute for Plant Pathology, investigations of *Trichoderma* biocontrol efficacy were conducted at INRA (Paris, France), Institute for Biological Control (Darmstadt, Germany) and in USA at Cornel University (Geneva, NY) Dr. Harman.

Another most famous and useful *T. harzianum* strain was T22 (also known as 1295-22, KRL-AG2, and ATCC 20847). It was produced by Dr. Harman in 1980s and was licensed from Cornell University by the Eastman Kodak Company, which developed the toxicity package and environmental studies and make registration possible. In about 1990, Kodak decided to abandon the agricultural pesticide market and gifted the registration of T22 and other data they had generated to Dr. Harman and his colleagues at the Cornell Research Foundation who founded a company, now BioWorks Inc. This company was founded by Dr. Harman and two of his colleagues previously under name TGT Inc. as their efforts were to develop biocontrol systems for commercial agriculture and to translate biocontrol research into biocontrol reality [14]. They encapsulated T22 in commercial products RootShield and T22 Planter Box and marketed under new company BioWorks Inc. (Geneva, NY, USA). Sales of those products began in 1993, and in 1998, sales have increased for 20% per year where it was marketing internationally with special consideration for its limited shelf life. Strain T22 was generally promoter of plant growth as well as mycofungicide against soil-borne plant pathogens. In plant, strain T22 enhances expression of proteins involved in photosynthesis and starch accumulation, and supposing its effects are due to increased photosynthetic rates in infected plants [8]. Interestingly, this strain was produced using protoplast fusion in order to be highly rhizosphere competent that also possess substantial ability to compete with spermosphere bacteria. Strains that were fused were *T. harzianum* T-95 and T-12 because first was a rhizosphere competent mutant produced from a strain isolated from a *Rhizoctonia*-suppressive Colombian soil, and second was more capable of competing with spermosphere bacteria than T-95 under iron-limiting conditions [14, 15]. The novel generation of RootShield is based on two *Trichoderma* species and contains strain *T. virens* G-41 together with strain *T. harzianum* T-22. Commercial name of product is RootShield PLUS and some with instruction that should be applied to disease-free plants previously chemically treated with fungicide, because the aggressively growing strains T-22 and G-41 are growing on the outside of roots and do not enter the plant tissue.

The availability and dispersion of *Trichoderma*-based biofertilizers are more widespread then commonly known with the tendency to expand due to the easier registrations. Mostly permitted for use in organic farming in Europe is: RootShield, Plant Box and Bio Trek (northern Europe, USA), Binap (Switzerland, Sweden, UK, USA), Bio fungus (Belgium), Supersivit (Czech Republic), Trichodex (Italy), Trifender (Hungary), and Trianum (Avantagro, Spain) [16]. Novel *Trichoderma* formulations are not based on single culture of one species but come as consortia or mix of at least two or three species or different strains of same species. Compatible consortia of compatible strains with different mechanisms, disease suppressive, or plant growth promoting, which complementary each other were found to be more effective than the application of individual organisms. Mixtures for biocontrol have a broadened range of pathogens against which are effective. Mixtures in plant growth promoters are based on insight that metabolic products of various *Trichoderma* strains are not identical and they have selective character to different plant species and even a variety [17, 18]. Seems this may be due to better interaction of some *Trichoderma* species or some strains with certain plant species because root exudates may induce or inhibit their mycelial growth [19]. In development are other types of consortia, mixtures of *Trichoderma* strains with other organisms, fungi or bacteria that are known as bioagent also. Because knowing of action mechanisms allows combining of strains with different modes of action in order to anticipate efficacy of final product, investigations of effectiveness differences between species and biotypes of same species are in progress.

3. *Trichoderma* as opportunistic plant symbiont

Mycoparasitic and nematophagous *Trichoderma* species found their prey in rhizosphere where roots are resort of plant pathogenic fungi and nematodes. Therefore, it becomes usual to define species of genus *Trichoderma* as free living rhizosphere organisms which colonize plant root surface as opportunistic plant symbionts. Biocontrol of plant pathogens by *Trichoderma* was from the first Weindling report in 1932 [1] considered as the direct ability of these fungi to interact with soil pathogens. Along with revelation of diverse antifungal mechanisms of

Trichoderma, the ability to promote plant growth and to increase plant height, leaf area, and dry weight were perceived. Positive influence of *Trichoderma* to a faster germination and increase in percentage of emergency were perceived also.

Abundant healthy roots are from environment where *Trichoderma* grows and proliferates best. The sucrose that leaks from roots stimulates growth of *Trichoderma* mycelium and leads to interaction with plant. In order to stimulate plant to provide more sucrose, *Trichoderma* has evolved numerous mechanisms for better routing. With its enzyme arsenal, *Trichoderma* enhances solubility of soil nutrients which will be otherwise unavailable to plant. Further, *Trichoderma* enhances nutrient uptake by plant and better plant nourishment that will result in stronger routing which are frequently associated with increase in yield [20]. Special benefit is induction of increased nitrogen use efficiency in plants. This mechanism is not enlightened yet but probably is connected to *Trichoderma* stimulation of deeper rooting, and thereby increasing the volume of soil colonized by plant roots. Plants approximately take up only 33% of the amount of applied nitrogen fertilizer while some field trial data on *Trichoderma* treatments in several different crops indicate possibility of reduction nitrogen application rates by 30–50% with no reduction in yield [21, 22]. Calculation says that if this reduction was applied to the 30 million hectares of wheat in the USA, the savings in nitrogen application would total more than a billion of nitrogen kilos annually.

Sucrose metabolism increased by *Trichoderma* stimulates the resistance response in the leaves that leads to increased photosynthesis and respiration because growth induced by *Trichoderma* plant requires energy, and sunlight energy utilized in increased photosynthesis will be the energy source needed for plant growth enhancement by *Trichoderma* mechanisms. Of course, better photosynthesis enables that more sucrose is translocating to the roots and metabolic circle continues. Increased leaf mass enables increased photosynthesis but the *Trichoderma* has the abilities to increase photosynthetic efficiency [21, 23]. It was demonstrated that electron flow was substantially increased by root colonization and electron transport strongly enhanced [24]. In trial with barley exposed to water deficiency *T. harzianum* substantially increases water deficit tolerance and consequently reduces effects on photosynthetic systems even when plants were at or approaching the permanent wilting point after 2 weeks of withholding irrigation. Therefore, plants benefit from relationship with *Trichoderma* through increased root, shoot, and leaf growth and increased macro- and micronutrient uptake and disease protection as well.

First scientific papers about effects of *Trichoderma* on plant growth promotion, mostly of horticultural crops and conifers, began to appear in 1980s and continuing in 1990s [18, 25–28]. There is ample documentation for the *Trichoderma* influence on plants which are colonized with effective *Trichoderma* strain and that they are substantially different from an uncolonize plants in quality and quantity of yield, withstand to adverse environmental conditions and pathogen attack. Positive influence of *Trichoderma* to a faster germination and increase in percentage of emergency were perceived also [17, 18, 29].

In Croatia, investigations with autochthon *Trichoderma* strains and their influence to growth of fiber flax, lettuce, tomato, cabbage, and red beet were conducted [16, 30–32]. Significant increase of some lettuce quality characteristics was gained with *T. viride* strain TPS applied in the form of alginate-pellets in two frequently used commercial potting compost mixture: Klasmann-Deilmann P 002 (Germany) and Stender A240 (Germany). Except dry weight, TPS enhanced the formation of leaves: at Stender difference against control varies for 1–2 leaves more, at Klasmann for 1 leaf more. Leaf length was longer for 2 cm at Stender and Klasmann amended with TPS-pellets than at control, while leaf width was wider for 3.15 cm at Stender and for 4.27 cm at Klasmann. Fresh weight was greater for 5.36 g at Stender and for 4.68 g at Klasmann against control. Dry weight was only characteristic on which TPS pellets did not have significant influence perhaps due to the similar nutrient content of Stedman and Klasmann substrates. These substrates are characterized by the use of fine peat with the addition of nutrient specially designed to meet the needs of young plants, so they similar in nutrient content (N 150–260 mg l^{-1}, P 180–280 mg l^{-1}, K 200–350 mg l^{-1}, and Mg 80–150 mg l^{-1}). *Trichoderma* is able to solubilize nutrients but only the ones present in substrate, and as Klasmann and Stender are enriched with the similar nutrients, there were no significant differences among them as trial variants. The differences were bespeaking when the TPS pellets were added against control. In Croatian trial with cabbage and red beet, the indigenous *T. viride* isolates STP16 and STP8 enhanced plant growth in only one trial vegetation season, and results confirm the hypothesis that biotypes of same species differ in their abilities to induce plant growth, so that growth promotion of *Trichoderma* is not species dependent as well as that biotypes of same species differ in their abilities for inducing plant growth. Influence of strains STP16 and STP8 on cabbage growth was estimated by weighing the heads. Fresh weight was greater at STP16 treatment (FW = 1666.5 g) than at STP8 treatment (FW = 1372.5 g) but not statistically different although in comparison to control (FW = 1291 g) significantly increased. Dry weight was slightly but statistically significantly increased at STP16 treatment (DW = 8.2 g) against STP8 treatment (DW = 7.2 g) which was statistically equal to control (DW = 6.3 g). Growth promotion index showed that STP16 treatment promotes fresh weight for 29% while STP8 treatment only for 6.3% and dry weight for 30.16% while STP8 for 14.29%. Influence of those strains on red beet growth was estimated by weighing the root. Fresh weight was increased by both isolates, STP16 (FW = 725 g) and STP8 (FW = 607.5 g). There was no statistically significant difference between isolate influences although STP16 significantly increased fresh weight in comparison to control (FW = 569 g), while STP8 did not. Dry weight was greater at STP16 treatment (DW = 13.1 g) and statistically significant in comparison to STP8 treatment (DW = 12.2 g) and control (DW = 11.6 g). Growth promotion index showed that STP16 treatment increased root fresh weight for 27.42%, while STP8 treatment for only 6.44%. Index calculated for dry weight showed that STP16 increased dry weight for 12.93% and STP8 for 5.17%. In trial with fiber flax indigenous *T. harzianum* strain STP, applied in the form of alginate-pellets and through pelleted seeds, positively influenced germination, seedling emergence and plant growth. Seedling emergence from pelleted seeds were delayed, and on 7th day, after seeding, 16% were emerged, while at STP-pellet treatment, 66% emerged and at control 53%. In the presence of strain STP, plants grow higher (66 cm) than on control (62 cm) and the highest where plants from pelleted seeds (72 cm).

First, this ability was treated as side effect of suppression of plant pathogenic fungi [25, 33–35]. Other possible explanations of this phenomenon include: control of minor pathogens leading to stronger root growth and nutrient uptake [26], secretion of plant growth regulatory factors such as phytohormones [33, 34, 36, 37], and release of soil nutrients and minerals by increased saprophytic activity of *Trichoderma* in the soil. Removing of a toxic and inhibitory material to plant growth from the soil was also presented as interesting explanation of *T. harzianum* plant growth promotion [25].

Today, *Trichoderma* positive effect on plant growth is considered as independent ability and equally remarkable and significant as its antifungal ability because growth enhancement has been observed in the absence of any detectable disease and in sterile soil [17, 28, 31]. Novel genetically analysis shown that the most of *Trichoderma* biocontrol activity is through their abilities to induce plant defense mechanism described as systemic disease resistance [21]. For example, antibiosis of *T. virens* against

Rhizoctonia solani on cotton seedlings and mycoparasitism *T. harzianum* on *Pythium* were found to be due solely to induced resistance [22, 38]. Therefore, the intimate *Trichoderma*-plant relationship is what induces localized and systemic resistance plant responses to pathogen attack, promoting plant growth and support/encourage biocontrol. From now on, the meaning of biological control must be expanded to include induction of plant defense mechanism up to the disease resistance and plant growth promotion along with classical antagonism like antibiosis and mycoparasitism.

Symbiotic *Trichoderma*-plant relationship results with effects that extend beyond biocontrol because *Trichoderma* directly influence plant physiology. That is why it is considered that they establish pseudomycorrhizal relationship with plant host. Species of *Trichoderma* are actually strong plant invaders and can colonize plant internally in endophytic manner with the ability to grow with plants. These allow much longer periods of efficacy than nonendophytic organisms and provide benefits to plants for at least the life of an annual crop [8].

Trichoderma species penetrate into the root cortex, epidermis, and a few cell layers below this level, based upon evidence with *Trichoderma* mutant strains that produce green fluorescent protein and electron microscopy and therefore are similar to endomycorrhizal fungi. Small protein from hydrophobin group is produced on the surface of *Trichoderma* hyphae and facilitates their attachment to root. For penetrations, *Trichoderma* uses appressoria which coil about root hairs and are similar to those observed in mycoparasitism. The *Trichoderma* enables further its entry with help of small protein swolenin TasSwo that recognize cellulose and modify plant cell structure. Entering the cells they have access to plant nutrients, which allows them to proliferate.

Penetration of *Trichoderma* hyphae into plant tissue is infection similar to one of other fungal plant pathogens but does not incite parasitism even though they have enzyme systems fully capable of macerating plant tissue. Although rarely phytopathogenic, one case of *T. virens* pathogenic on cotton seedlings was described and enlightened. It happened because protein responsible for antibiotic production, a single 18 kDa, that induces resistance was not expressed, so resistance was not induced [39, 40]. This nonpathogenic yet plant beneficial life style is a successful strategy for the fungus because it provides *Trichoderma* with more sucrose from enlarged roots due to better plant nourishment and also prey for mycoparasitic and nematophagous strains. Recently, endophytic *Trichoderma* strains are known which colonize vascular systems of certain plants [39] like it is known for other root colonizing biocontrol fungi such as binucleate *Rhizoctonia* and nonpathogenic *Fusarium* species [41, 42]. It was discovered that cocoa permits *Trichoderma* strain ramification throughout their structure. Mostly, plants will not permit that so the same strain will function only as root colonists applied to other plants [43].

After penetration into the root tissue, *Trichoderma* establishes chemical communication with the plant and interact at molecular level. Inside the root cells, *Trichoderma* can modify plant's gene expression to activate its immune system. These result initially in an induction of resistance mechanisms, so plant form thickened cell walls and produce phenolic depositions that intercept the *Trichoderma* to the area of infection and prevent further plant colonization [39, 44]. This is the type of plant localized resistance to *Trichoderma*. Therefore, disease development does not occur and this is asymptomatic infection but plant defense system is triggered.

Plants have two immune systems: basal disease resistance is systemic acquired resistance (SAR) and the other is induced systemic resistance (ISR). SAR is induced by pathogens and follows the salicylic acid pathway to reduce the severity of pathogenicity, while ISR follows jasmonic acid pathway. The first plant response to *Trichoderma* infection was found to increase in salicylic and jasmonic acid levels and typical antipathogenic peroxidase activity. Infected plant cells recognize that they are under pathogen attack by detecting pathogen-associated molecular patterns (PAMPs), also called microbe-associated molecular patterns (MAMPs) which is more suitable to use for *Trichoderma.* This molecular pattern is essential for the pathogen life which is binding to pattern recognition receptors (PPT) on cell surface, and this triggers basic immunity system SAR. PAMPs are not found in plant, so thus the plant's receptor recognized them as being potentially dangerous and this triggers the plant SAR [45]. Simultaneously, infected plant cell recognizes pathogen through toxins considered being effectors, and this triggers plant's effector-triggered immunity (ETI). ISR is triggered by infection of beneficial microorganisms such as *Trichoderma*, recognized by chitin presence and induced by jasmonic acid and ethylene by wound signals, which they transmit from roots to other plant's part. When *Trichoderma* penetrates root, the ISR is induced by pattern recognition receptors localized in the plasma membrane of plant cell. First response is hypersensitive reaction resulting from production of antimicrobial compounds intent to restrict/restrain the potential pathogen. Plant can identify *Trichoderma* by the following PAMPs: cellulases, chitinases, endopolygalacturonase, peptaibols, and 6-pentyl-α-pyrone [46]. More *Trichoderma* biocontrol compounds that are able to induce plant defenses and are connected with plant-beneficial effects are different proteins, ceratoplatanins, polygalacturonase, cellulose-binding-domain proteins, and nonactive xylanase, and secondary metabolites, peptaibiotics, pyrones, pyridines, and butenolides, The first discovered chemical communicator of *Trichoderma* was a 22 kDa xylanase, protein which induces localized resistance in plants [39]. The peptides or proteins that are effective have masses of 6.5, 18, 20, 32, and 42 kDa. Finding that many of them retained their activity as denatured proteins was lead to premise that particular amino acid sequences are the important factor in their activity rather than enzymatic function.

Once when this *Trichoderma*-root biochemical cross-talk begins, *Trichoderma* may influence plant response to other pathogens attack by increasing SAR and ETI [47]. Effective biocontrol strains of *Trichoderma* changes in the amplitude of plant ETI by using the zig-zag model proposed by Jones and Dangle [48]. *Trichoderma* strains that activate both ISR and SAR or only SAR were also known. In investigation of cucumber root colonization by *Trichoderma* applied at high concentrations, 28 proteins whose expression was affected were identified in cotyledons [49]. All were regulated by *Trichoderma*, and among them, 17 were found to be upregulated, while 11 were downregulated. Proteins differentially regulated by *Trichoderma* were involved in isoprenoid and ethylene biosynthesis, and in metabolism of photosynthesis, photorespiration, and carbohydrate. Important finding is that *Trichoderma* can influence plant's oxidizing, reactive chemical species containing oxygen (ROS). They are a natural by-product of the normal metabolism of oxygen which is important in cell signaling and homeostasis. ROS are influenced by plant responses to stress, mostly to abiotic environmental conditions, so their levels can drastically increase under stress. They have high energy and are unstable so easily react with other species, like biomolecules such as DNA, and oxidizing them. This is known as oxidative stress and may result in significant damage to cell structures. Involvement of *Trichoderma* in ROS scavenging enlightens why *Trichoderma* treated plants are better coping the stress by drought or salinity. One of the pathways is the glutathione-ascorbate cycle, and *Trichoderma* enhancement of enzymes in it will recycle antioxidants more rapidly and thereby reduce stresses effects. In trial with barley exposed to water deficiency, *T. harzianum* substantially increases water deficit tolerance and consequently reduces effects on photosynthetic systems even when plants were at or approaching the permanent wilting point after 2 weeks of withholding irrigation. That shows possible *Trichoderma* influence on photosynthesis. Plants enhanced by *Trichoderma* have increased leaf mass that enables increased photosynthesis, but *Trichoderma* also have the abilities to increase photosynthetic efficiency [21, 23]. It was demonstrated that electron flow was substantially increased by root colonization and electron transport strongly enhanced [43].

Some *Trichoderma* strains can counteract pathogen toxins (effectors) in two ways: inhibiting pathogenicity factors and influencing pathogen dispersal and nutrition, whereby the ETI is induced as mentioned earlier. *Trichoderma* can improve ETI by releasing compounds that plant receptors will recognize as pathogen effectors or causing faster response. Other strains can induce stronger immunologic plant response than pathogen, meaning they induce ISR more than pathogen induces SAR, by producing a variety of MAMPs such as hydrophobins, expansin-like proteins, secondary metabolites and enzymes having direct antimicrobial activity. There is interesting example of involvement of *T. arundinacea* in induction of plant defense. Its trichothecene toxin harzianum A has been recognized as MAMPs in tomato seedlings and activated both ISR and SAR and primed them against *B. cinerea* and *R. solani* [50]. Trichothecenes are produced by a number of fungal genera like *Fusarium*, *Stachybotrys*, and *Myrothecium*. So far, *Trichoderma* is the only trichothecene producer in which the tri5 gene is not located in the main tri cluster. *Trichoderma* species produces two trichothecene toxins: harzianum A and trichothecene.

Evidently, *Trichoderma* antifungal activity against phytopathogenic fungi and nematodes, as well as competition for nutrients conferring a nutritional advantage, are carried out by production of extracellular hydrolytic enzymes and/or the production of secondary metabolites with antifungal activity. It is postulated that the *Trichoderma* ability to act as soil colonizer, pseudomycorrhizal relationship with plant host with well being for plant health and fitness, mycoparasite and nematofag are to be provided by *Trichoderma* genome. Genome coevolution has been demonstrated in many plant-pathogen interactions, so it can be considered in the case of some plant and *Trichoderma* species. Strong genetic components to the responses of at last maize to *T. harzianum* T22 were confirmed in trials with a series of inbred lines which were preceded by large trial conducted in U.S. corn-belt with 160 maize hybrids and T22 [44]. Today, several hundred separate plant genes or proteins are known whose expression is altered by *Trichoderma* root colonization although the expression consequences are more pronounce in the shoots than in the roots [22]. Considering all described, it is evident that proteomics study is need to give an understanding of how *Trichoderma*-treated plants become more resistant to pathogen attacks.

4. Applying *Trichoderma* in agriculture

Scientific articles describing excellent antifungal *Trichoderma* efficacy against some phytopathogenic fungi or enhancement of plant growth and yield are published worldwide each year. Many originate from countries having economics primarily based on agriculture and describe mostly experimental results achieved by native strain or created product based on it and may not be applicable on large-scale crop production. Furthermore, novel articles summarize biocontrol programs, models of commercialization, and registration requirements. Dominantly is emphasizing that full-scale production, marketing, and registration requirements are unfavorable for products of biocontrol agents and simply too expensive, especially for the agricultural-based countries in which they are needed the most. All those articles clearly show that isolation, experimenting to evaluate biocontrol activity of strain and encapsulating it in formulations for low-scale trial needs are achievable part of developing biocontrol product. To be able to overcome issues on product industrialization and commercial and registration requirements, *Trichoderma* research community depends on stakeholders for investments in that part of developing process and industrial linkages. That's why the use of *Trichoderma*-products in developing countries falls under local production model for using microbial agents. This model is one of the four economic models for using microbial agents proposed by Hartman et al. [8].

First model is the microbial pesticide model which implies full registration of microbial product as pesticide and marketing worldwide. It is established in developed countries in USA, Canada, and EU; although in USA, it is substantially different. Registration of microbial products as pesticides is based on the interpretation of the term "pesticide." It does not necessarily refer to killing (e.g., fungicide) or inhibiting (e.g., fungistatic) pest only that it is controlled. Therefore, *Trichoderma*-products fall into the scope of pesticide although registration requirements will be substantially less than those for synthetic chemicals, but remain difficult. Good example for this model is T22-product, and data that approximately $12 million were required for registration, development of product facility, formulation, and marketing system before its sales began to grow. In USA, using the microbial pesticide model requires a minimum of $8 million and 3–6 years before highly effective product is established in the market. In Canada in EU, registration regulatory also requires efficacy evaluation with toxicological and environmental testing, while in USA, they are required over time. Efficacy tests are required for almost every crop-pathogen combination and for *Trichoderma*-based products with broad capabilities on many crops and pathogens; this limits or even precludes registration and even makes it almost impossible from a financial standpoint. On the other hand, marketing them as plant-growth enhancements and strengthening agents gives them a market advantage because registration requirements like time and efficacy tests on pathogens are excluded. Although, in the USA, mycorrhizal fungi and rhizobia are not subject to regulatory approval for use while in EU and Canada are needed. Need to overcome the expensive efficacy evaluation of *Trichoderma*-products led to creation of model named *Inoculants, plant strengthening agents and biofertilizers*. This model is used in various agricultural systems where *Trichoderma* products are marketed as plant inoculants for improvement of plant performance, but the pesticidial claims are not made although their diseases control benefits and well known. This gives those product marketplace advantages because many necessities for registration are avoided and therefore takes less time to reach the marketplace. Sales of *Trichoderma*-products may be larger than the sales of registered products. Great example is the product based on strain T22 which was registered as biofungicide in USA but in EU was just the beginning registration process in 2010. Until then, it was sold as plant strengthening agent named Triannum although its biofungicidal activities were known. As reason for delayed EU registration product, authors instigate economic because that the return on investment for full European registration was unlike to occur. In comparison to microbial pesticide model, the plant inoculants model leak two steps required for registration. Both models have followed steps: identification of good agent; development of production and formulation system; patenting of strain and/or process; building large-scale production and nationwide or international marketing, but steps: toxicology and other testing and registration are required only in microbial pesticide model. Local production model has only one step—discovery of good strain. As name tells, the production of autochthon strain is local. Strains are grown and multiplied in order with well-known methods for semi-solid cultivation on wheat or corn bran, rice or similar substrate or in liquid fermentation. Cultivation and growing the required amount of inoculum is timed to be delivered directly before the application, date can be ordered by the grower, what eliminates extensive production and formulating. This model describes production of plan growth promotion rhizobacteria (PGPR) and *Trichoderma*-formulation at Faculty of Agriculture University of Zagreb in Croatia. The last one is in Croatia produced only by this chapter author. Although it is regulated by the government, local production model is different from model named *Governmental monopolies or state-supported production*. This model is based on Cuban rapidly shift from conventional agriculture to semi-organic farming. After collapse of trade relations in 1989, Cuba sets in economic and food crisis. As the pesticide and fertilizer imports were reduced to more than 80%, they established alternative agricultural technology and urban agriculture (composed of about 8000 gardens nationwide) with biological pest control practices supported with production on biopesticides and biofertilizers on a large scale. During 1996, Havana's urban farms provided the city's urban population with 8500 tons of agricultural produce, 4 million dozens of flowers, 7.5 million eggs, and 3650 tons of meat. This averted the crisis as it was said in Ref. [8, 51, 52] "It thus helps refute the most common argument—that we couldn't "feed the hungry" without pesticides—against taking the "ethical" position in real-world pest management policy debates." In Venezuela, a complete on large-scale implementation of biopesticide/biofertilizer development program has been funded and run through the Instituto Nacional de Investigaciones Agricola. In India, government is promoting all steps connected to adoption of biopesticides through various types of legislative, like the National Farmer Policy.

4.1 Assay for biological activity

Members of genus *Trichoderma* are among the most prevalent cultivable fungi in soils, based upon the frequency of isolation on suitable media. They are present in all types of temperate and tropical soils and constituted up to 3% of the total fungal propagules (mycobiota) in forest and 1.5% in pasture soils, and their populations range from 10^2 to 10^3 per gram of soil. Therefore, *Trichoderma* strains are mostly isolated from soil, dominantly forest soil, and organic substrates form rhizosphere: alive, dead, healthy, or diseased plant tissue (root, green parts), common on wood decaying and fungal structures (fruiting bodies, sclerotia, and mycelial mats) [53]. Mentioned sclerotia and mycelial mats can be used as bates in search for antagonist for specific fungal pathogen. For that purpose, those mycelial structures are burying in natural soil sample. In looking for strain for biocontrol, it is recommended to identify the problem, the target pathogen (soil borne or aerial, source of inoculum, biology, and epidemiology) and its niche, host crop, and environment conditions. The most effective strains are sought in the geographic center of plant origin because as pathogens are coevolving with the host, their antagonist coevolving also. Strain can be isolated from stromata of teleomorph *Hypocrea* which are often found on wood and less frequently on some Basidiomycetous fungi (sedges, bracket). Isolating process is not inexpensive, requires time, and needs labor. When isolating from various sources in nature, one must have in mind that it is easy to obtain mixed cultures of *Trichoderma* species, as well as teleomorphic *Hypocrea* state, because they usually intermingle. To proceed investigation with pure culture, it is necessary to grow culture from single spore, conidia, chlamydospores, or better ascospores if it is possible and even hyphal tips. This can be achieved using dilution series of soils, root, and other plant tissue macerates suspensions of fungal structures which can be then plated on potato dextrose agar (PDA), corn meal agar (CMA) or *Trichoderma* medium E (TME) or other selective methods have been devised by numbered authors. For example, there is selective media developed to distinguish "P" strains of *T. virens* that produce the antibiotic gliovirin and are effective against *Pythium*, than "Q" strains that produce gliotoxin and are effective against *Rhizoctonia*. Cultivation of strain and multiplication for further tests is not difficult, and the methods described by Rifai (1969) are generally still followed. Recommended are inoculation on oatmeal, cornmeal or malt extract agars, and incubation under daylight for 5–7 days at a temperature of 20–25°C as they allow observation of stable morphological features [54–58]. Less expensive isolation is achieved when strain occurs naturally and is isolated from an area where they will be used after semi-solid cultivation, so it is connected to local production model. Hence, just isolation is not expensive but finding the useful strains in evaluating process is.

After isolation, bioassay for biological activity will be required to determine their nature, whether it is suppressing pathogens or enhancing plant growth, and if it is performed satisfactory. Moreover, the molecular identification of species is required so that the harmful (mushroom pathogens) and dangerous (trichothecene producers) ones can be avoided. Serious mushroom diseases can cause *T. aggressivum*, *T. pleurotum*, and *T. pleuroticola*. This species are genetically distinct from well known biocontrol strains [8, 59, 60]. Trichothecene production and their role in induction of plant defense were discussed earlier but it needs to be emphasizing that for registration and marketing strain must be nontoxic as microbial pesticides model requires toxicity testing. But, as most strains will be marketed under other production model, the potential for harm cannot be avoided entirely. Toxicity information is available for five *Trichoderma* species. Some testing reported acute oral toxicity as being >500 to <2.000 mg/kg, so at highest level, no effects were seen, and further, there have been no reported reactions after many years of extensive use. The species in *T. brevicompactum* complex are trichothecene producers, and they are as well not related to *Trichoderma* strains that are registered as bioagent. Immunosuppressive mycotoxin gliotoxin is produced by *T. virens* "Q" strains earlier mentioned, but they were no reported mycotoxicosis attributed to *T. virens* like it were for *Aspergillus* spp. Antibiotic important for biocontrol is volatile lactone and pyrone as it inhibits spore germination of *Phytophthora*, *Botrytis*, and other fungi. It is produced by *T. viride/H. rufa* clade only. Pyrone has pleasant coconut odor and is present in fruits, and it is used as flavoring agent, and therefore not considered as high hazard. In connection to plant defense system, peptaibols were mentioned earlier also. As they can lyses red blood cells that can be potentially harmful but it was reported that the syntheses are induced only by fungal cell walls or other elicitors, so again are connected with *Trichoderma* antifungal activity.

Considering all, finding a biologically useful and perspective strain for encapsulation into successful formulation which will be commercially viable is hard. *Trichoderma* has advantages because it can be easily isolated, grown, and tested for selection of efficient strains, manipulated and encapsulated in various formulations as they have good shelf life, which aids commercialization. Most strains used in bioapplications were local in origin and were used locally or regionally. Only smaller number of strains will show to be useful in various locations and environmental conditions, and therefore they become widely adopted and commercially available, like famous T22 and T39. So, what are the characteristics of perspective *Trichoderma* strain? One of the most important traits of beneficial culture is rhizosphere competence, the ability to survive in the environment, mostly rhizosphere. Rhizosphere competence or competitive saprobic ability is culture ability to compete other microbes in colonizing cellulose-rich substrates or the intercellular spaces of the surface layers of host roots. Therefore, *Trichoderma* culture would require cellulolytic enzymes to occupy this region and interact with the plant at a molecular level, either for plant growth promotion or inducing defense mechanism and inhibiting pathogenicity [57, 61]. Rhizosphere competence is always associated to cellulose production, and this is used for assessment of culture ability to metabolize cellulose. Method is simple as some cellulose materials (straw, cellophane discs) are buried into filed soil, then inoculate with a known quantity of fungal propagule (conidia, chlamydospores) and after incubation reisolate. The amount of cellulose produced is directly related to the culture rhizosphere competence.

After isolation of strain with high rhizosphere and competitive saprophytic competence which is easily artificially mass multiplicative and safe for the environment, it must be perspective by its broad spectrum of activity. It can provide excellent and reliable control against a more of pathogens or/and can enhance growth of different plant species. The presentation must be of low effective doses. Classical method *in vitro* for assessment of biocontrol efficiency is "dual-culture" method. Probably the oldest one dual culture technique was described in 1955 by Morton and Strouble [62]. In one Petri plate, usually 10 mm diameter one, pathogen and *Trichoderma* are confronted as antagonist in the form of mycelial disc cut from the margin of the 7 days-old culture with a circular cutter. The mycelial disc is 5 mm in diameter but there are variations, so it can be 10 cm in diameter [63]. Both discs are placed on the substrate on the opposite of the plate, usually PDA, in Petri plate in the manner that the mycelium side facing the substrate and gently pressed in. The interspace between them amounted to 5 cm and interspace between micellar discs and edge of plate 2 cm. In control plate, *Trichoderma* disc is replaced with sterile agar disc. Inoculated plates are incubated at 25 ± 1°C for 7 days usually, although it could be 10–14 days depending on the pathogen species. Recommendable is to do readings on 2, 4, and 6 days after the incubation period by measuring radial growth of pathogen. The inhibition is quantified with index of inhibition (I) which represents inhibition percent of average pathogen radial growth (T) in the presence of *Trichoderma* and is calculated in relation to growth of the controls (C) as follows: $I(\%) = [(C - T)/C] \times 100$ [64]. Basic mechanisms can be determined: competition for substrate and nutrients as antagonist grow faster and colonize substrate; mycoparasitism if antagonist colonize pathogen mycelial mat, sclerotia or fruiting body and diminish and decay of those structure can be monitored; antibiosis when inhibition zone is formed because toxins spread through substrate and inhibited growth of pathogen. There is possibility to misinterpret mycoparasitic capability of the antagonist if neglected that it can absorb nutrients from the agar. Antibiosis by production of volatile metabolites can be tested using slight modification of dual culture technique [6]. The mycelial discs are placed centrally, but the *Trichoderma* disc is placed in the lid while pathogen disc on the bottom of the plate sealed together with adhesive paraffin tape.

Furthermore, perspective culture needs to be compatible with other bioagents or tolerant to pesticides commonly used in agriculture which facilitates integrated control. For instance, it is recommended by commercial seed treatment systems to apply a chemical pesticide with the effective *Trichoderma* strains. While chemical will provide short term protection, the *Trichoderma* provides season-long benefits to plants as colonizes roots. Culture should even tolerate oxidizing agent, UV radiations, desiccation, heat, draught, etc. [21, 57, 65, 66]. Because these characteristics are strain depended, fewer than 1% of screened cultures will meet the expectations. Therefore, screening for bioactivity is the critical step when hundreds or thousands of cultures are tested. First, efficacy bioassays are performed in small-scale laboratory trials, *in vitro* and *in vivo,* which are usually random. Recommended is so called three-partner model, a system with potential *Trichoderma* agent, pathogen, and plant. Screening process is obviously time and money consuming as well as requires labor of more than one analyst. Molecular techniques of assisted selection using phenotypic or/and genotypic markers perhaps can improve screening productivity and shorten it but will raise the costs. Moreover, only in few cases, connection with bioactivity and molecular markers was proved. Few perspective cultures from lab-trials are proceeding to greenhouse and small pilot trials at open. At this stage, testing is conducted in different environmental conditions and crops and several pathogens in the case of biocontrol activity. Satisfaction of these very limiting conditions is important for future possible wide range of applicability of bioagent which is the most important property needed for registration and marketing. For commercialization, the bioagent should be produced on industrial scale.

4.2 Methods of formulation

Trichoderma is mostly fermented in solid state with the aim to achieve highest yield with lower cost of culture medium. Obtained fungal biomass needs to be immobilized in certain carriers in low cost but high density inocula and encapsulated into formulation engineered to maintain fungal propagule viable during the transport, storage, and application. *Trichoderma* has advantages because it can be easily manipulated for encapsulation which means mixing wet or dry fungal biomass with a matrix forming material, such as gelatinized polysaccharide or an oil emulsion. Matrix is serving as carrier of fungal inoculums. Most of the examples on different types of *Trichoderma* inoculum for biocontrol include peat, granular vermiculite or clay mixtures, grains, and alginate pellets. Encapsulation in formulation of alginate pellets has been studied and positively evaluated by various authors as found to be successful for the *Trichoderma* delivery [30–32, 67–72]. In the receipt of chapter's author for small trial purpose, culture is grown on Petri dishes 10 cm in diameter containing 20 ml of PDA and incubated in humid chamber at 25°C for 7 days until conidiation. After incubation, the substrate altogether with hyphal biomass and conidia from two Petri dishes are upraised with spatula and transferred into glass with 50 ml sterile DI water. These are mixed by common blender at low speed for 3–5 min in order to make a suspension. The final concentration to be used contained 4×10^6 spores ml^{-1}. The suspension is mixed with 100 g l^{-1} talcum and 10 g l^{-1} sodium alginate. The formed matrix is placed in a separator funnel modified in order to allow suspension to drip into a 0.1 M suspension of calcium gluconate under stirring on magnetic agitator. Drops of alginate matrix dripped into calcium gluconate suspension transformed to gelatinized spherules or pellets. Pellets were removed from suspension within 10 min, rinsed with distilled water, and allowed to dry on waxed paper under a sterile vertical flow for 12–24 h [16, 30, 31]. In India is quite popular a talc-based formulation of *Trichoderma* developed at Tamil Nadu Agricultural University [64]. There are also oil-based formulations prepared with a combination of vegetable/mineral oils and they are suitable for foliar spraying under dry weather. Some formulations use organic wastes like coffee husk from coffee industry and press mud, byproduct of sugar factory.

Carriers of *Trichoderma* inoculum must be cheap, should dissolve well in water, and preserve fungal viability to insure formulation shelf-life. Formulation good for the commercialization should have increased shelf-life to fulfill requirements for storage and transport, and this is one of the most limiting factors. Further, it should deliver viable propagules in adequate concentration through adequate application. In the case of endophytic strains that grow with plant roots, only small amounts of inoculum are needed to be provided for long-term benefits. The minimal propagule number should not be less than 2×10^6 per milliliter or gram of formulation. Commercial preparations are mostly high concentrated with more than of 10^{10} propagules per gram. Thus, commercial preparations need to be applied in dose of 500 mg per hectare or for addition to greenhouse potting soils only 10^4–10^5 per cm^3 [21].

The most important act before commercialization is that formulation, or even strain/culture should be legally protected by means of patent. It is patented as biotechnological invention because it is based on microbe and its mechanisms or its metabolic products. The strain pure culture needs to be deposited in an officially recognized microbial collection by Budapest treaty signed by all countries pertaining to the World Intellectual Property Organization. It should be emphasized that patent is not authorization for commercial use, and it does not connote registration for agricultural use. All patents are regulated by legal treaties. There are series of international and national treaties: Union of Paris of 1883, Patent Cooperation Treaty of 1970, and European Patent Agreement of 1973. In the USA is United States Patents while in EU the legal framework is regulated by Directive 98/44CEE on the patentability of biotechnological inventions.

5. Concluding remarks

Present pest control intend to kill, even eradicate organisms but without choosing between harmful and benefit ones. Biocontrol could help in providing low-cost and environmentally safe technologies to farmers especially today when food security and rural livelihood are a key priority. Original paper updates on basic and applied research in all aspects of biological control of invertebrate, vertebrate and weed pests, and plant diseases can be found in the BioControl, the official journal of the International Organization for Biological Control (IOBC). For developing world, biopesticides and biofertilizer are considered extremely important as perceived by Association of Asian Pacific Agricultural Research Institution (AAPARI). Thus are in progress biopesticide researches of the agriculturally important microorganisms led by the *Trichoderma* that can be encapsulated in bioproducts. Although the chemical control of plant diseases differs tremendously from biocontrol by microbial-based biopreparations, registration regulations remain the same or similar depending on country. In USA, the Environmental Protection Agency (EPA) registers pesticides under the Federal Insecticide, Fungicide and Rodenticide Act (FIFRA). In EU, registration is defined by Regulatives 91/414/CEE, 2092/91, 1488/97, etc. In Croatia, it is defined by Plant Protection Law, which is harmonized with EU Regulatives. In general, registration of any product making a pesticidal claim is obligation defined by law in order to prevent unreasonable, adverse effects on consumer health, or the environment. Registration requires time, expenses, and efforts for conducting toxicological, environmental, and efficiency testing. Especially is difficult to enlighten excessive specificity of bioagent as success of control by any bioagent depends on three living system: pathogen, plant, and bioagent.

In this new era, the meaning of biological control must be expanded to include plant growth promotion and disease resistance along with classical antibiosis and mycoparasitism because *Trichoderma* biocontrol and plant performances-enhancing activities overlap. Even better, *Trichoderma* research community emphasizes that considering biocontrol as primary ability of *Trichoderma* may influence biocontrol system in development because it means optimizing conditions for wrong mechanism. It must be taking into account that *Trichoderma*, as pseudomycorrhizal plant partner, has effects that extend beyond biocontrol. These fungi produce changes in plant metabolism, which have direct effects on plant physiology, like increasing growth and enhancing resistance, and even reprogramming of plant gene expression owing to coevolution with plants. Thus is especially pronounced that the control of plan response to abiotic and biotic stresses, such as diseases, is only a subset of their activities and benefits to plant. In future can be expected momentum of proteome study in order to help giving an understanding of how *Trichoderma* treated plants become more resistant to pathogen attacks. Also, needed is development of easy and inexpensive screening methods whose test conditions approach as much as possible the real system where biocontrol has to be. Registration requirements need to be revised and include fact that *Trichoderma*-fungicides are entering market as plant growth promoters. There are opinions that future of biopesticides lies in plant-protecting pesticides or self-protecting plant, a high-value crop plant with embedded genes from bioagent. Yet, are not the same concerns influencing transgenic plant and biopesticide in the light of biosafety? Nontarget effects, toxicity, and possible pathogenicity for plant, animals and humans, allergenicity or horizontal gene transfer to nontarget organisms are exactly limiting issues for both transgenic plant and biopesticide. Further, for both groups are present consumer concerns about living microorganisms in connection with bioterrorism and food-bone diseases. Noticeable is that all that is needed is socially receptive environment and that should be developed and promoted by *Trichoderma* research community.

Author details

Snježana Topolovec-Pintarić
Department of Plant Pathology, University of Zagreb Faculty of Agriculture, Zagreb, Croatia

*Address all correspondence to: tpintaric@agr.hr

References

[1] Weindling R. *Trichoderma lignorum* as a parasite of other soil fungi. Phytopathology. 1932;**22**:8372-8451

[2] Weindling R. Studies on a lethal principle effective in the parasitic action of *Trichoderma lingorum* on *Rhizoctonia solani* and other soil fungi. Phytopathology. 1934;**24**:1153-1179

[3] Dennis C, Webster J. Antagonistic properties of species groups of *Trichoderma* I, production of non-volatile antibiotics. Transactions of the British Mycological Society. 1971;**57**:25-39

[4] Weindling R. Studies on a lethal principle effective in the parasitic action of *Trichoderma lignorum* on *Rhizoctonia solani* and other soil fungi. Phytopathology. 1934;**24**:1153-1179

[5] Weindling R. Experimental consideration of the mold toxins of *Gliocladium* and *Trichoderma*. Phytopathology. 1941;**31**:991-1003

[6] Dennis C, Webster J. Antagonistic properties of species-groups of *Trichoderma*: II. Production of volatile antibiotics. Transactions of the British Mycological Society. 1971;**57**:41-48. DOI: 10.1016/S0007-1536(71)80078-5

[7] Woo SL, Ruocco M, Vinale F, Nigro M, Marra R, Lombardi N, et al. *Trichoderma*-based products and their widespread use in agriculture. The Open Mycology Journal. 2014;**8**:71-126. DOI: 10.1007/s00253-016-7792-1

[8] Harman GE, Ma O, Samules GJ, Lorto M. Changing models for commercialization and implementation of biocontrol in the developing and the developed world. Plant Disease. 2010;**94**(8):928-939. DOI: 10.1094/PDIS-94-8-0928

[9] Sharma P, Sharma M, Raja M, Shanmugam V. Status of *Trichoderma* research in India: A review. Indian Phytopathology. 2014;**67**:1-119

[10] Elad Y, Gessler C, Pertot I. Integrated pest management—Italian Israeli cooperation in research and development. In: Proceedings of the Italian-Israeli Workshop on Agriculture, Research and Cooperation: 2003; Israel. Business Conference. 14-17 December 2005; Hong Kong. New York: IEEE; 2006. pp. 866-870

[11] O'Neill TM, Elad Y, Shtienberg D, Cohen A. Control of grapevine grey mould with *Trichoderma harzianum* T39. Biocontrol Science and Technology. 1996;**6**:139-146. DOI: 10.1080/09583159650039340

[12] Topolovec-Pintarić S, Cvjetković B, Jurjević Z. Experience in integrated chemical-biological control of grey mould (*Botrytis cinerea*) on grapevines in Croatia. Journal of Wine Research. 1999;**10**(1):33-41. DOI: 10.1080/09571269908718156

[13] Topolovec-Pintarić S, Cvjetković B. The sensitivity of *Botrytis cinerea* Pers:Fr. to pyrimethanil in Croatia. Journal of Plant Diseases and Protection. 2002;**109**(1):74-79. DOI: 10.1007/BF03356289

[14] Harman GE. Myths and dogmas of biocontrol. Plant Disease. 2000;**84**(4):377-393. DOI: 10.1094/PDIS.2000.84.4.377

[15] Ahmad JS, Baker R. Rhizosphere competence of *Trichoderma harzianum*. Phytopathology. 1987;**77**:182-189. DOI: 10.1094/Phyto-7-182

[16] Topolovec-Pintarić S, Žutić I, Đemić E. Enhanced growth of cabbage and red beet by *Trichoderma viride*. Acta Agriculturae Slovenica. 2013;**101**(1):87-92. DOI: 10.2478/acas-2013-0010

[17] Celar F, Valic N. Effects of *Trichoderma* spp. and *Gliocladium roseum* culture filtrates on seed germination of vegetables and maize. Journal of Plant Diseases and Protection. 2005;**112**:343-350

[18] Gupta O, Sharma ND. Effect of fungal metabolites on seed germination and root length of black gram. (*Phaseolus mungo* L.). Legume Research. 1995;**18**:64-66

[19] Bal U, Altinatas S. Effects of *Trichoderma harzianum* on lettuce in protected cultivation. Journal of Central European Agriculture. 2008;**9**(1):63-70

[20] Altomare C, Norvell WA, Björkman T, Harman GE. Solubilization of phosphates and micronutrients by the plant—Growth—Promoting and biocontrol fungus *Trichoderma harzianum* Rifai. Applied and Environmental Microbiology. 1999;**65**:2926-2933

[21] Harman GE. *Trichoderma*—Not just for biocontrol anymore. Phytoparasitica. 2011;**39**:103-108. DOI: 10.1007/s12600-011-0151-y

[22] Shoresh M, Harman GE, Mastouri F. Induced systemic resistance and plant responses to fungal biocontrol agents. Annual Review of Phytopathology. 2010;**48**:21-23. DOI: 10.1146/annurev-phyto-073009-114450

[23] Vargas WA, Mandawe JC, Kenerley CM. Plant-derived sucrose is a key element in the symbiotic association between *Trichoderma virens* and maize plants. Plant Physiology. 2009;**151**:792-808. DOI: 10.1104/pp.109.141291

[24] Rai MK, Shende S, Strasser RJ. JIP test for fast fluorescence transients as a rapid and sensitive technique in assessing the effectiveness of arbuscular mycorrhizal fungi in *Zea mays*: Analysis of chlorophyll a fluorescence. Plant Biosystems. 2008;**142**:191-198. DOI: 10.1080/11263500802150225

[25] Ousley MA, Lynch JM, Whipps JM. Potential of *Trichoderma* spp. as consistent plant growth stimulators. Biology and Fertility of Soils. 1994;**17**:85-90. DOI: 10.1007/BF00337738

[26] Ousley MA, Lynch JM, Whipps JM. Effect of *Trichoderma* on plant growth: A balance between inhibition and growth promotion. Microbial Ecology. 1993;**26**:277-285. DOI: 10.1007/BF00176959

[27] Lynch JM, Wilson KL, Ousley MA, Whipps JM. Response of lettuce to *Trichoderma* treatment. Letters in Applied Mitrobiology. 1991;**12**:59-61. DOI: 10.1111/j.1472-765X.1991.tb00503.x

[28] Kleifeld O, Chet I. *Trichoderma harzianum*—Interaction with plants and effect on growth response. Plant and Soil. 1992;**144**:267-272. DOI: 10.1007/BF00012884

[29] Koch E. Effect of biocontrol agents on plant growth in the absence of pathogens. IOBC/WPRS Bulletin. 2001;**24**(1):81-89

[30] Topolovec-Pintarić S. Influence of *Trichoderma harzianum* Rifai on the fiber flax germination and growth. Növénytermelés. 2010;**59**(4):421-424. DOI: 10.1556/Novenyterm.59.2010.Suppl.4

[31] Topolovec-Pintarić S, Žutić I, Lončarić I. Enhancing plant growth by *Trichoderma viride* based pellets. Növénytermelés. 2011;**60**(S1):177-180. DOI: 10.1556/Novenyterm.60.2011.Suppl.1

[32] Topolovec-Pintarić S, Vinceković M, Jalšenjak N, Martinko K, Žutić I, Đermić E. Prototype of tomato biofertilizer: *Trichoderma viride* and calcium based

microcapsules. In: Proceedings of the 52nd Croatian and 12th International Symposium on Agriculture; 12-17 February 2017; Dubrovnik. Osijek: Faculty of Agriculture University of J.J. Strossmayer; 2017. pp. 100-107

[33] Baker R. *Trichoderma* spp. as plant stimulants. Critical Reviews in Biotechnology. 1988;7:97-106. DOI: 10.3109/07388558809150724

[34] Chang Y-C, Chang Y-C, Baker R, Kleifeld O, Chet I. Increased growth of plants in presence of the biological control agent *Trichoderma harzianum*. Plant Disease. 1986;**70**:145-148. DOI: 10.1094/PD-70-145

[35] Inbar J, Abramsky M, Cohen D, Chet I. Plant growth enhancement and disease control by *Trichoderma harzianum* in vegetable seedlings growth under commercial conditions. European Journal of Plant Pathology. 1994;**100**:337-346. DOI: 10.1007/BF01876444

[36] Windham MT, Elad Y, Baker R. A mechanism for increased plant growth induced by *Trichoderma* spp. Phytopathology. 1986;**6**:518-521. DOI: 10.1094/Phyto-76-518

[37] Ousley MA, Lynch JM, Whipps JM. The effects of addition of *Trichoderma* inocula on flowering and shoot growth of bedding plants. Scientia Horticulturae. 1994;**59**:147-155. DOI: 10.1016/0304-4238(94)90081-7

[38] Howell CR. Understanding the mechanisms employed by *Trichoderma virens* to effect biological control of cotton diseases. Phytopathology. 2006;**96**:178-180. DOI: 10.1094/PHYTO-96-0178

[39] Harman GE, Shoresh M. The mechanisms and applications of opportunistic plant symbionts. In: Vurro M, Gressel J, editors. Novel Biotechnologies for Biocontrol Agent Enhancement and Management. NATO Security Through Science Series. Dordrecht: Springer; 2007. pp. 131-155. DOI: 10.1007/978-1-4020-5799-1_7

[40] Howell CR. Mechanisms employed by *Trichoderma* species in the biological control of plant diseases: The history and evolution of current concepts. Plant Disease. 2003;**87**:4-10. DOI: 10.1094/PDIS.2003.87.1.4

[41] Benhamou N, Garand C, Goulet A. Ability of nonpathogenic *Fusarium oxysporum* strain Fo47 to induce resistance against *Pythium ultimum* infection in cucumber. Applied and Environmental Microbiology. 2002;**68**:4044-4060. DOI: 10.1128/AEM.68.8.4044-4060.2002

[42] Hwang J, Benson DM. Expression of induced resistance in poinsettia cuttings against *Rhizoctonia* stem rot by treatment of stock plants with binucleate *Rhizoctonia*. Biological Control. 2003;**27**:73-80. DOI: 10.1016/S1049-9644(02)00185-8

[43] Bae H, Roberts DP, Lim HS, Strem M, Park SC, Ryu CM, et al. Endophytic *Trichoderma* isolates from tropical environments delay disease and induce resistance against *Phytophthora capsici* in hot pepper using multiple mechanisms. Molecular Plant-Microbe Interactions. 2011;**24**:336-351. DOI: 10.1094/MPMI-09-10-0221

[44] Harman GE, Howell CR, Viterbo A, Chet I, Lorito M. *Trichoderma* species—Opportunistic, avirulent plant symbionts. Nature Reviews. Microbiology. 2004;**2**(1):43-56. DOI: 10.1038/nrmicro797

[45] Bittel P, Robatzek S. Microbe-associated molecular patterns (MAMPs) probe plant immunity. Current Opinion in Plant Biology. 2007;**10**:335-341. DOI: 10.1016/j.pbi.2007.04.021

[46] Hermosa R, Viterbo A, Chet I, Monte E. Plant-beneficial effect of *Trichoderma* and its genes. Microbiology. 2012;**158**:17-25. DOI: 10.1099/mic.0.052274-0

[47] Lorito M, Woo SL, Harman GE, Monte E. *Trichoderma*: From 'Omics to the field. Annual Review of Phytopathology. 2010;**48**:395-417. DOI: 10.1146/annurev-phyto-073009-114314

[48] Jones J, Dangl J. The plant immune system. Nature. 2006;**444**:323-329. DOI: 10.1038/nature05286

[49] Seggara G, Casanova E, Bellido D, Odena MA, Oliveira E, Trillas I. Proteome, salicylic acid, and jasmonic acid changes cucumber plants inoculated with *Trichoderma asperellum* T34. Proteomics. 2007;7:3943-3952. DOI: 10.1002/pmic.200700173

[50] Malmierca MG, Cardoza RE, Alexander NJ, McCormic SP, Hermosa R, Monte E, et al. Involvement of *Trichoderma* trichotecens in biocontrol activity and induction of plant defense-related genes. Applied and Environmental Microbiology. 2012;**78**:4856-4868. DOI: 10.1128/AEM.00385-12

[51] Altieri MA, Companioni N, Cañizares K, Murphy C, Rosset P, Bourque M, et al. The greening of the "barrios": Urban agriculture for food security in Cuba. Agriculture and Human Values. 1999;**16**(2):131-140. DOI: 10.1023/a:1007545304561

[52] Rosset PM. Cuba: Ethics, biological control, and crisis. Agriculture and Human Values. 1997;**14**(3):291-302. DOI: 10.1023/a:1007433501248

[53] Chet I. Innovative Approaches to Plant Disease Control. 2nd ed. New York: Wiley & Sons; 1987. p. 372. ISBN: 0471809624 9780471809623

[54] Kubicek CP, Harman GE, editors. *Trichoderma* and *Gliocladium*: Basic Biology, Taxonomy and Genetics. 1st ed. London: Taylor & Francis e-Library; 1998. p. 278. ISBN: 0748405720

[55] Elad Y, Chet I. Improved selective media for isolation of *Trichoderma* spp. or *Fusarium* spp. Phytoparasitica. 1983;**11**:55-58. DOI: 10.1007/BF02980712

[56] Howell CR. Selective isolation from soil and separation *in vitro* of P and Q strains of *Trichoderma virens* with differentiated media. Mycologia. 1999;**91**:930-934. DOI: 10.2307/3761624

[57] Samuels GJ, Hebbar PK. *Trichoderma*: Identification and Agriculural Applications. 1st ed. St. Paul: The American Phytopathological Society; 2015. p. 196. ISBN-13: 9780890544846

[58] Rifai MA. A Revision of the Genus *Trichoderma*. 1st ed. Kew: Commonwealth Mycological Institute; 1969. p. 56. ISBN: 0851990002 9780851990002

[59] Samuels GJ, Dodd SL, Gams W, Castlebury LA, Petrini O. *Trichoderma* species associated with the green mould epidemic of commercially grown *Agaricus bisporus*. Mycologia. 2002;**94**:146-170. DOI: 10.2307/3761854

[60] Kredics L, Kocsube S, Nagy L, Komon-Zelazowska M, Manczinger L, Sajben E, et al. Molecular identification of *Trichoderma* species associated with *Pleurotus ostreatus* and natural substrates of the oyster mushroom. FEMS Microbiology Letters. 2009;**300**:58-67. DOI: 10.1111/j.1574-6968.2009.01765.x

[61] Baker R. Diversity in biological control. Crop Protection. 1991;**10**:85-94. DOI: 10.1016/0261-2194(91)90054-U

[62] Morton DT, Stroube NH. Antagonistic and stimulatory effects of

microorganism upon *Sclerotium rolfsii*. Phytopathology. 1955;**45**:419-420

[63] Porras M, Barrau C, Santos B, Arroyo FT, Blanco C, Romero F. Effects of temperature on in vitro response of *Trichoderma* strains against strawberry pathogen *Rhizoctonia solani* Kühn. Plant Protection Science. 2002;**38**(2):620-622

[64] Edington LV, Khew KL, Barron GI. Fungitoxic spectrum of benzimidazole compounds. Phytopathology. 1971;**61**:42-44. DOI: 10.1094/Phyto-61-42

[65] Montesinos E. Development, registration and commercialization of microbial pesticides for plant protection. International Microbiology. 2003;**6**:245-252. DOI: 10.1007/s10123-003-0144-x

[66] Kumar S, Thakur M, Rani A. *Trichoderma*: Mass production, formulation, quality control, delivery and its scope in commercialization in India for the management of plant diseases. African Journal of Agricultural Research. 2014;**9**(53):3838-3852. DOI: 10.5897/AJAR2014. 9061

[67] Walker HL, Connick WJ. Sodium alginate for production and formulation of mycoherbicides. Weed Science. 1983;**31**:333-338. Available from: https://www.jstor.org/stable/4043716

[68] Leštan D, Leštan M, Chapelle JA, Lamar RT. Biological potential of fungal inocula for bioaugmentation of contaminated soils. Journal of Industrial Microbiology. 1996;**16**:286-294. DOI: 10.1007/BF01570036

[69] Mafia RG, Alfenas AC, Maffia LA, Ventura GM, Sanfuentes EA. Encapsulamento de *Trichoderma inhamatum* para controle biologico de *Rhizoctonia solani* na propagacaoclonal de Eucalyptus. Fitopatologia Brasileira. 2003;**28**:101-105. DOI: 10.1590/S0100-41582003000100016

[70] Nipoti P, Manzali D, Gennari S, D'Ercole N, Rivas F. Activity of *Trichoderma harzianum* Rifai on germination of asparagus seeds. Acta Horticulturae. 1990;**271**:403-407. DOI: 10.17660/ActaHortic.1990.271.57

[71] Knudsen GR, Bin L. Effects of temperature, soil moisture, and wheat bran on growth of *Trichoderma harzianum* from alginate pellets. Phytopathology. 1990;**80**:724-727. DOI: 10.1094/Phyto-80-724

[72] Lewis JA, Papavizas GC. Characteristics of alginate pellets formulated with *Trichoderma* and *Gliocladium* and their effect on the proliferation of the fungi in soil. Plant Pathology. 1985;**34**:571-577. DOI: 10.1111/j.1365-3059.1985.tb01409.x

Section 2

Trichoderma as Biological Control Agent

Chapter 3

Trichoderma as a Biocontrol Agent against *Sclerotinia* Stem Rot or White Mold on Soybeans in Brazil: Usage and Technology

Fernando C. Juliatti, Anakely A. Rezende, Breno Cezar Marinho Juliatti and Tâmara P. Morais

Abstract

Biological control agents are alternatives to chemical pesticides in the management of plant diseases. Currently, hundreds of bioproducts are commercially available in international market varying mainly in antagonistic microorganisms and formulation. We screened four *Trichoderma*-based products as to their efficacy in controlling *Sclerotinia* stem rot (SSR) under protected and field environments and their effect on soybean seeds' sanity and physiological qualities. We also tested application technologies through seed microbiolization and foliar spraying to deliver the microorganisms, and their compatibility with chemical fungicides. In vitro assays showed that all Trichoderma strains were antagonistic to *S. sclerotiorum* evidencing hyperparasitic activity. Moreover, the bioproducts reduced fungi incidence on soybean seeds, promoted faster seedling emergence and did not hamper seeds' vigor. Increases of 14 and 37% were registered for root length and shoot fresh weight over that of the untreated control indicating potential application of the bioproducts as soybean growth promoters. Thiophanate-methyl and procymidone were the most compatible, without drastically affecting spore germination or mycelium growth. Under field conditions, all *Trichoderma* strains reduced SSR incidence and increased soybean grain yield. Formulation interferes on bioproducts' viability and efficacy deserving special attention upon development.

Keywords: biological control, *Sclerotinia sclerotiorum*, *Glycine max*, hyperparasitism, physiological seed quality

1. Situation of *Sclerotinia* stem rot in Brazil

The white mold or Sclerotinia stem rot (*Sclerotinia sclerotiorum*) is an important disease in Brazil and in the fields is founded more than 7 million ha with the disease on soybeans in populations of the sclerotia between 1 and 500 per meter square (**Figure 1**). The evolution of agricultural practices over the years undoubtedly increased global food production [1, 2]. Those practices included the use of fertilizers, machinery, improved genetic plant materials, adoption of different cropping systems (no-tillage, crop rotation, intercropping), and intensive use of chemicals. In this scenario, pests coevolved with crops demanding new management strategies,

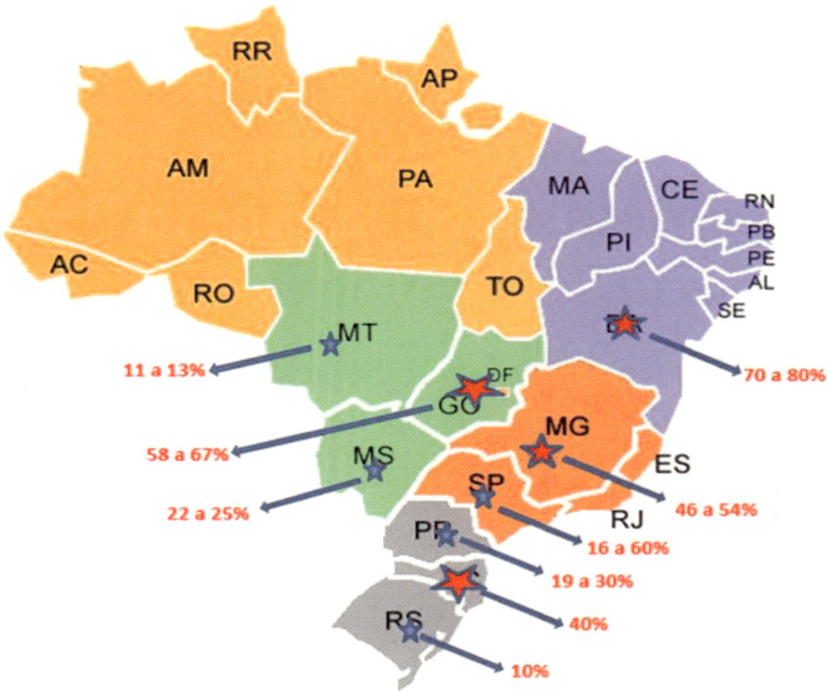

Figure 1.
Estimation of the production area in Brazil (soybean, beans, and cotton), in which S. sclerotiorum had been detected in crops (1–40% incidence in plants) [sources: researchers from universities, foundations, rural extension agencies, cooperatives, and consultants, through personal communication]. Red star—higher levels of incidence, blue star—medium to lower levels of incidence. Sclerotinia affects an estimated area of 7.5 million ha (22.9%). Total area to soybean production in Brazil is 33.9 million ha (CONAB and IBGE—Brazil).

as diseases and insects' outbreaks became recurrent [3–8]. Chemical pesticides have long been used in plant protection. However, they frequently increase production costs, negatively affect the environment, and are ineffective against resistant populations [9, 10]. The search for eco-friendly efficient alternatives to control plant diseases is not a recent need. One option includes the use of antagonistic microorganisms, such as *Trichoderma* spp.

Trichoderma is a genus of soil-borne fungi with well-known anti-phytopathogen activities. Mechanisms of action include competition, mycoparasitism, antibiosis, and host-induced systemic resistance [11]. *Trichoderma* species compete with pathogens mainly for nutrients and ecological niches. Besides rapid growth and abundant production of spores, some strains can synthesize siderophores that inhibit the growth of other fungi during competition [12]. Mycoparasitism traits of *Trichoderma* rely on activity of cell wall-degrading enzymes secreted by the fungus after its hyphae coil around and further penetrate the pathogen's hyphae cells [13].

Trichoderma also produces various antimicrobial compounds that can be purified and directly used against pathogenic fungi [14–16]. The last mechanism of action triggers systemic defense responses in host plants, which interfere with pathogen's establishment, colonization, and multiplication [14, 17]. In addition to the mentioned mechanisms, some *Trichoderma* strains promote plants protection by stimulating their development [18, 19].

The first report on the use of *Trichoderma* as a biological control agent of phytopathogens in Brazil dates back to 1950 against tobacco mosaic virus [20]. Further research evaluated its potential controlling plant pathogenic fungi and oomycetes that led to the development of *Trichoderma*-based commercial products [21]. Among them, bioproducts recommended for the control of the etiologic agent of stem rot disease [*Sclerotinia sclerotiorum* (Lib.) de Bary] deserve special attention [22]. *Sclerotinia* stem rot (SSR) is one of the most relevant diseases in soybean crops [*Glycine max* (L.) Merrill]. Early symptoms are water-soaked irregular spots that progress to brownish lesions and eventual necrosis, wilt of leaves, and plant death. A characteristic white cotton-like mycelium on infected tissues is diagnostically detectable. At later stage of disease development, survival structures of the fungus, named sclerotia, are formed on hosts. Seed lots contaminated with fungus mycelia or sclerotia constitute the most common source of dissemination of the pathogen. Under favorable climate conditions, the disease can impair soybean yields in up to 70% [22].

Efficient chemical control of SSR relies on prophylactic application of fungicides, since curative spraying does not revert yield losses despite being effective in reducing the inoculum potential for subsequent crops. The intensive long-term fungicide-based management strategies for the control of this disease resulted in the development of resistant *S. sclerotiorum* strains toward many active ingredients (such as carbendazim, dimetachlone, and thiophanate-methyl), demanding constant baseline sensitivity studies to monitor the field efficacy of chemicals [23–29]. The need to introduce alternative molecules to control this devastating pathogen, allied to environmental and food securities, opened the market for bioproducts. Worldwide, many entrepreneurs invest in the development of formulations and registration of bioproducts. Commercialization, however, is often hindered by farmer's mistrust on products' consistent performance under field conditions, as most research are confined to the laboratory and monitoring of quality is not regularly done [30].

We accessed the efficacy of different *Trichoderma*-based fungicides on the control of SSR and their possible effect on initial development of soybean plants. The bioproducts differed in formulation and strains and were both tested in vitro and under field conditions. We also verified the application technology, through spraying or seed treatment, and compatibility with chemical fungicides, as those are important aspects to address further applications of bioproducts in agriculture by demonstrating consortium to other plant disease management strategies.

2. Development of biological control in Brazil using *Trichoderma*

2.1 Microorganisms and growth conditions

Trichoderma strains used in this research were recovered from bioproducts available on the market (**Table 1**) after plating in PDA medium (20% potato extract, 2% dextrose, and 2% agar). All inhibition assays were against the *S. sclerotiorum* strain Jatai, characterized as highly aggressive [31]. This strain was field-isolated from a commercial soybean crop at the state of Goias, Brazil. Pathogen was recovered in PDA medium from sclerotia previously disinfested with 50% (v v^{-1}) ethanol for 30 s followed by immersion in 0.5% (w v^{-1}) sodium hypochlorite solution for 1 min. Later, sclerotia were rinsed thrice with sterile distilled water and incubated for myceliogenic germination. Microorganisms' incubation conditions were at 22 ± 2°C under 12 h photoperiod, otherwise stated, for indicated periods.

2.2 Monitoring of quality of bioproducts

Products (**Tables 1** and **2**) were serially diluted in sterile distilled water and plated in PDA medium for 5 days. *Trichoderma* titer was obtained by counting the

Bioproduct codification	Active ingredient/ microorganism	Formulation	Titer reported by the manufacturer
SF04	*T. asperellum*[a]	WG	1.0×10^{10}[e]
IBLF006	*T. harzianum*[b]	WP	5.0×10^{10}[e]
ESALQ-1306	*T. harzianum*[c]	CS	2.0×10^{9}[f]
Tricho	*Trichoderma* spp.[d]	WP	1.0×10^{8} [e]

WG, wettable granule; WP, wettable powder; CS, concentrated suspension

[a]*Strain SF04*

[b]*Strain IBLF006*

[c]*Strain ESALQ-1306*

[d]*Not specified on product's label*

[e]*Colony-forming units (CFU) m* L^{-1} *or CFU* g^{-1}

[f]*Viable conidia m* L^{-1}

Table 1.
Information on the bioproducts used in this study.

	SF04	IBLF006	ESALQ-1306	Tricho
Bioproduct label	1.0×10^{10}[a]	5.0×10^{10}[a]	2.0×10^{9}[b]	1.0×10^{8}[a]
Hemocytometer	8.8×10^{10}	1.2×10^{10}	1.9×10^{10}	6.7×10^{9}
Plate[a]	5.0×10^{10}	3.0×10^{6}	6.0×10^{9}	1.0×10^{7}
Bacterial contamination[c, d]	5.0×10^{5} A	6.0×10^{6} C	1.0×10^{6} A	3.0×10^{6} B
Viability (%)[c]	98 A	60 C	98 A	90 B

Titer of Trichoderma spp. was accessed through the number of viable spores counted in a hemocytometer slide and of colony-forming units recovered in PDA plates. Concentrations were compared to titers reported by the manufacturer. Bacterial contamination and spore viability also determine bioproducts' quality

[a]*CFU* mL^{-1} *or CFU* g^{-1}

[b]*Viable conidia* mL^{-1}

[c]*Averages followed by different uppercase letters are statistically different by the Tukey test* ($p < 0.05$)
[d]*Data transformed to* $\sqrt{x + 0.5}$

Table 2.
Monitoring of quality of bioproducts.

number of colonies grown in vitro and by the number of viable spores in a hemocytometer slide visualized under light microscope (Olympus CX40). The percentage of germinated spores was used to access fungus viability 24 h after incubation. Only spores with germ tube bigger than or equivalent to spore's own size were considered viable. Bacterial contaminations were expressed in colony-forming units (CFU) per mL or per gram of each bioproduct, detected 2 days after plating diluted aliquots of bioproducts in tryptic soya agar medium (1.5% casein peptone (pancreatic), 0.5% soya peptone, 0.5% sodium chloride, 1.5% agar, pH 7.3) [32].

2.3 In vitro antagonistic activity to *S. sclerotiorum*

Antagonism of *Trichoderma* strains to the phytopathogen *S. sclerotiorum* was verified by the dual culture assay modified from [33]. Briefly, 5-mm-dia mycelial agar discs of the fungi were collected from the edge of 3-day-old colonies and simultaneously placed on opposite sides of PDA plates. Plates were incubated as described [34]. Seven days after incubation, fungi growth was scored according to a modified scale proposed by Bell et al. [35]. We used quant software [36] to develop a diagrammatic scale in which *Trichoderma* strains were classified based on the area of the plate they covered. We considered antagonist or efficient those attributed scores up to 3.0 (at least 50% of the plate's surface). The search for efficient and environmentally safe alternatives to replace chemical pesticides routinely used in food production systems has brought to the market biological products containing antagonistic microorganisms. Among them, *Trichoderma* spp. deserves special attention. Surveys indicate the availability of more than 250 *Trichoderma*-based products in the international market and its growing adoption in Brazilian agriculture [37]. The diversity of products mainly comprises differences in fungus strains and formulations, aspects preponderant in bioproducts' efficacy, and shelf life. The quality of a biological product can vary from its manufacturing process to its application in the field. Therefore, it must be periodically checked to ensure efficiency. Although there is no standard methodology, the evaluation of the quality of a bioproduct is basically done by three criteria: concentration in plates, spore count, and viability. Verification of bacterial contamination is equally important as its presence reduces the shelf life of the product.

Higher concentrations of the antagonist in the tested bioproducts were recorded by direct quantification of the number of viable spores in a hemocytometer slide. Compared to quantification in a plate, this may occur because nearby propagules, after spread on plate's surface, visually form only one colony, underestimating the result, which does not happen in the individualized spore counting under a light microscope. From all counted spores, IBLF006 was the only bioproduct with germination percentage lower than 90, coincident to its higher contamination by bacterial cells. By accessing the concentration on the plate through the indirect counting method (serial dilutions of fungi suspensions), the bioproduct SF04 showed *Trichoderma* concentration beyond stated in the product's label, equivalent to 5.0×10^{10} CFU g^{-1} (**Table 4**). All evaluated bioproducts met the minimum concentration of the antagonist recommended for commercialization, of 2×10^{6} CFU mL^{-1} or CFU g^{-1} in culture medium, and no pathogenic contaminants belonging to the genera *Salmonella*, *Shigella*, or *Vibrio* [30]. However, they exceeded the bound for microbial contamination (1×10^{4} counts per mL/g).

There is a clear need to standardize and specify on the product's label the methodology adopted for quality monitoring, due to discrepancies between antagonist concentration values measured by spores counting (direct quantification) and by the number of in vitro colonies (indirect). Besides, limitations and modifications of the methodologies interfere in the result [38–41]. The maintenance of viability, especially during storage of the bioproduct, requires studies on formulations more adequate to the stability of microorganisms. In the market, *Trichoderma* is sold on its cultivation substrate, for example, bioproducts Tricho and IBLF006, which is ground and packed. This formulation (WP) hinders field spraying due to nozzle clogging and has less fungus viability (60–90%) and increased possibility of contamination by other fungi and bacteria (up to 12× higher than in the bioproduct SF04), reducing products' efficacy in the field. Other formulations available on the market, such as pure spores, spores mixed in oil, concentrated suspension (CS), or even wettable granules (WG), facilitate the application technology. To increase shelf life, adjuvants are usually added to protect propagules [42]. Recently, some bioproducts have evolved with the emergence of antimicrobial secondary metabolite formulations. Those formulations allow greater stability of the active ingredient under room temperature storage making it easier commercialization of the bioproduct without loss of quality. In addition, they offer advantages in terms of ease of application, protection against UV radiation under field conditions, action against the target pathogen without compromising the soil microbiota, and protection of the active ingredient when mixed to other chemicals [43]. It is very important to choose a product with quality combined with a formulation that guarantees stability and efficacy in the application and control of SSR.

The antagonistic effect of the strains was first verified through simultaneous cultivation under in vitro conditions, determining the area of the plate occupied by the colonies of *Trichoderma* and of *Sclerotinia*. According to the diagrammatic scale we developed, following the one proposed in 1982 [35], the bioproducts scored 2.5. Therefore, they did not differ statistically among themselves as to the percentage of the area of the plate they covered (62.3–64.4%) and were considered antagonist to *S. sclerotiorum*. Reduction of phytopathogen growth can be attributed to competition for space and nutrients of the culture medium, in which the great environmental fitness of *Trichoderma* and its rapid radial growth in in vitro cultivation are highlighted. In vitro dual culture assays are important tools in the selection process of biocontrol agents. Those tests provide useful information on strains' efficacy and variability and on pathogens' susceptibility to evaluated agents. The tests are conducted under controlled environment conditions minimizing variable effects of temperature, humidity, light, and soil microflora [15, 35]. We did not observe differences in antagonistic potency among the *Trichoderma* strains used in this study, for which the soybean-isolated *S. sclerotiorum* was susceptible. Dual culture assays are also used to analyze antagonist-phytopathogen interactions at ultrastructural level, with the aid of light or electron microscopy [44]. SEM images showed that all *Trichoderma* strains were able to colonize *S. sclerotiorum*, by either penetrating or strangling its hyphae (**Figure 2**). It also noted the growth of parallel hyphae [45]. Both interactions observed in our study, strangulation and penetration, can be interpreted as a hyperparasitic behavior of *Trichoderma* strains present in the bioproducts [46]. During hyperparasitism, *Trichoderma* species detect hyphae of

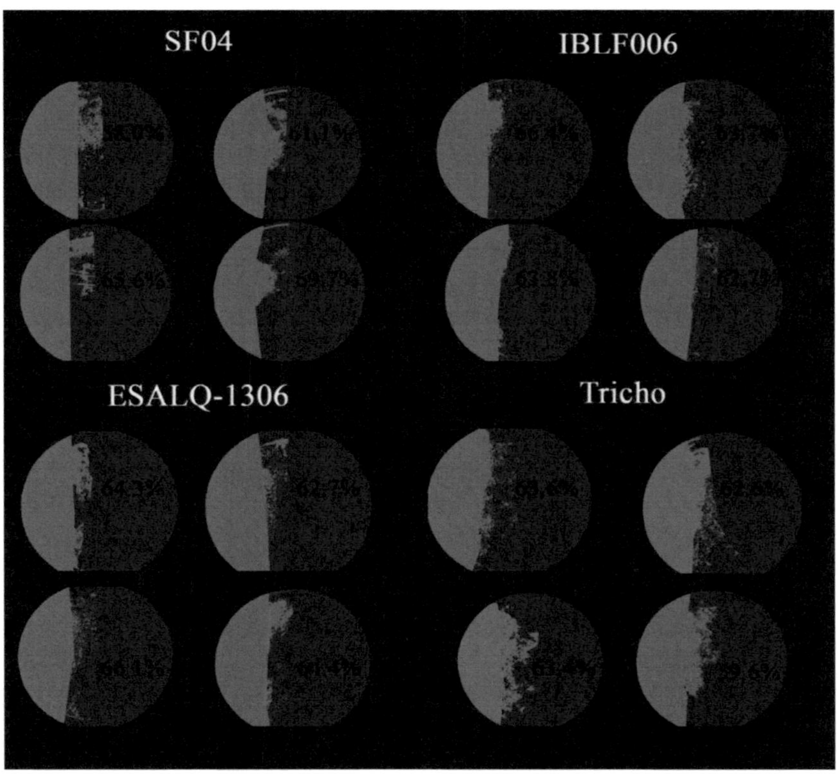

Figure 2.
Scanning electron photomicrography of the interactions between Trichoderma spp. and S. sclerotiorum showing strangulation (1) and penetration (2) of antagonist hyphae in the pathogen. Bioproducts: (a) SF04, (b) IBLF006, (c) ESALQ-1306, and (d) Tricho.

susceptible fungi due to chemical stimuli produced by the host hypha itself. The antagonists form appressory structures and tightly coil around the full extent of the hypha penetrating and degrading it [47–49]. This mechanism has already been demonstrated by many researchers through interactions of *T. harzianum*, *T. viride* and *T. asperellum* with *Rhizoctonia solani*, *Pythium* spp., *Fusarium oxysporum*, *F. solani*, *S. rolfsii*, *S. sclerotiorum*, *Botrytis cinerea*, *Phytophthora* spp., *Macrophomina phaseolina*, and *Alternaria solani* [45, 47–52].

The antagonistic activity of the bioproducts was also verified against the pathogen survival structures (data not shown). Percentage of non-germinated sclerotia and of sclerotia colonized by *Trichoderma* reached 55.6 and 71.25 (strain ESALQ-1306). Under simulated infested soil condition, treatments effectively controlled inoculum with a twofold reduction compared to the untreated soil. There was a negative correlation ($p < 0.01$, $r = -0.725$) between sclerotia germination and parasitism indicating that *Trichoderma* spp. colonization leads to reduction of the inoculum potential for subsequent crops. It should not be expected, however, that only direct ground spraying or seed treatment alone are effective in controlling *S. sclerotiorum*. The contact area of the antagonist with the soil is small limiting its proliferation. Besides, competition with the soil microbiota and possible adverse environmental conditions may compromise *Trichoderma* establishment. Still, direct ground spraying and seed treatment can be associated with other agricultural practices, such as mulching application of liquid compost and chemical fungicides [17, 53].

2.4 Hyperparasitism and antagonistic activity of *Trichoderma*-based products against *S. sclerotiorum* of *Trichoderma* spp. to *S. sclerotiorum*

To study the interaction between the antagonist and the pathogen, mycelia agar discs (5 mm diameter) from collation zones among both fungi colonies were collected at the seventh day of co-cultivation and further analyzed [54–55]. Discs were fixed to bristles in modified Karnovsky solution (2.5% glutaraldehyde and 2% paraformaldehyde in cacodylate buffer 0.05 M, pH 7.2), at 4°C for 17 h, followed by four rinses with mentioned buffer. Samples were subsequently fixed with 1% (w v^{-1}) osmium tetroxide in cacodylate buffer 0.01 M (pH 7.2) during 1 hour at 4°C, rinsed thrice with distilled water, and dehydrated in graded acetone series (30, 50, 70, 90, and 100%). Samples were kept at each solution for 10 minutes, and each step was repeated three times. Drying was done with carbon dioxide using a critical point dryer (TEC-030) (Balzers, Liechtenstein). Samples were fixed to aluminum stubs and gold-coated (20 nm/180 seconds) before visualized in a scanning electron microscope model LEO 435VP (Zeiss, Oberkochen). **Figures 2** and **3** show the hyperparasitism and antagonistic activity of Trichoderma-based products against *S. sclerotiorum* of *Trichoderma* spp. to *S. sclerotiorum* (different commercial products) used in Brazil.

2.5 Effect of *Trichoderma* spp. on sclerotia germination

Soil parasitism of *Trichoderma* strains against *S. sclerotiorum* was evaluated. Two-mm-dia sclerotia were buried at 0.5 cm depth in acrylic boxes containing autoclaved soil, which was later sprayed with the *Trichoderma*-based products following field doses recommended by the manufactures, at a spray volume of 100 mL. Sixteen sclerotia were evenly distributed per box. Mock consisted of application of sterile distilled water. After incubation for 5 days, we transferred the sclerotia to PDA plates and let them incubate for another 10 days. The number of germinated and parasitized sclerotia was counted, and the hyperparasitism action from Trichoderma species was reduced until 90% the viability of sclerotia from soil in the first year of treatment.

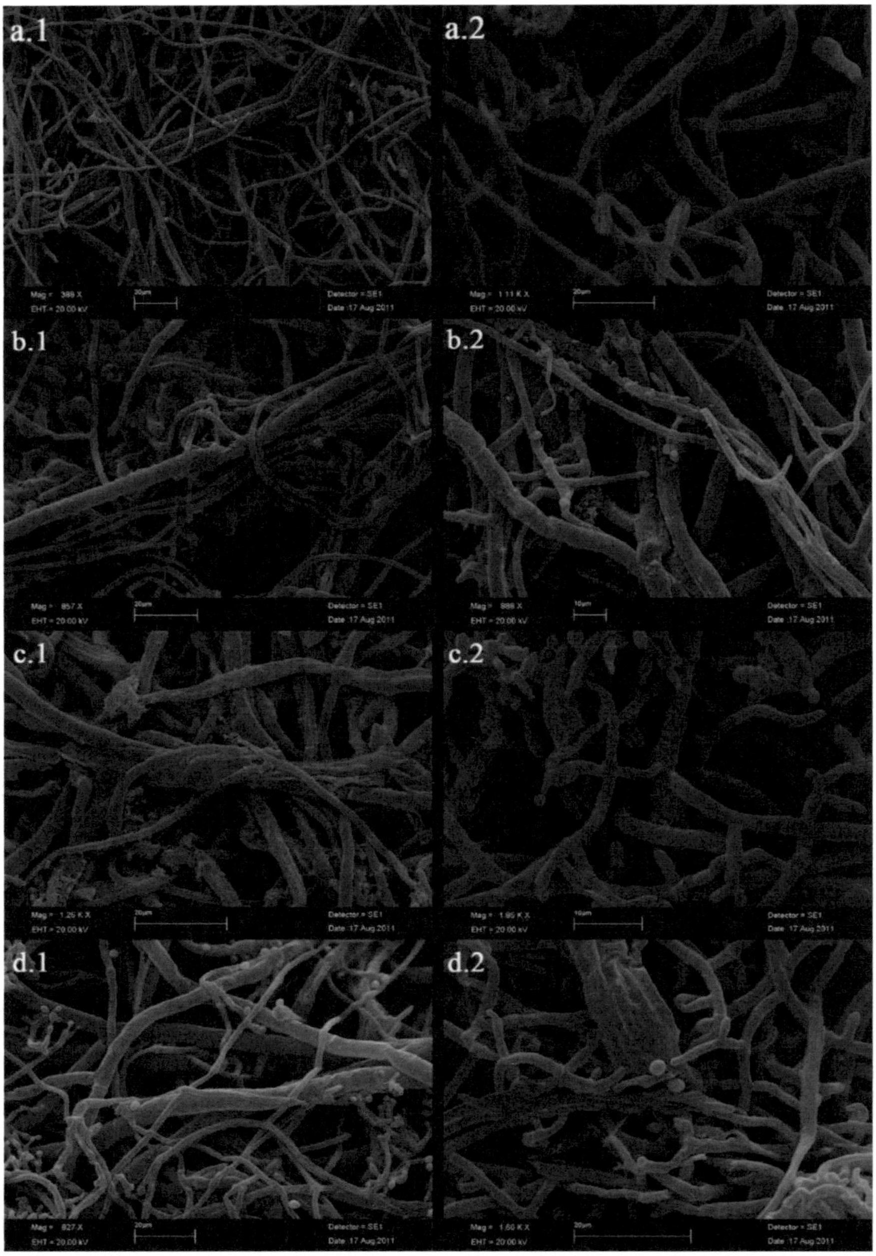

Figure 3.
Spore germination (%) of Trichoderma strains after treatment of soybean seeds with chemical fungicides and bioproducts. Viability was accessed right after seed treatment (ST), at 3 and 16 h after ST. Average of four replications (n = 200, each) ± standard deviations are shown. Different lowercase letters are statistically different by the Tukey test at p < 0.05.

2.6 Compatibility of fungicides with *Trichoderma* spp. in vitro and in seed treatment

Different fungicides (**Table 3**), usually used in the control of *S. sclerotiorum* under field conditions, were added to autoclaved PDA medium before polymerization (±40°C) to five final concentrations (0.1, 1, 10, 100 and 1000 ppm).

The medium was poured into Petri dishes and inoculated with 5-mm-dia colony discs of Trichoderma strains (**Table 4**). Plates were then incubated for 1 week. The diameter of Trichoderma colonies was measured daily. At the seventh day, we calculated the growth speed index (GSI) according to [57].

Seed treatment was done with soybean cultivar NK7074RR, considered susceptible to SSR [58]. Chemical and biological fungicides were applied at doses recommended by the manufactures (**Table 2**). Seeds were first treated with the chemical pesticides followed by application of the bioproducts. Positive controls consisted of seeds treated only with the *Trichoderma*-based products. Samples were collected at three different exposure times (0, 3, and 16 h after seed treatment). At each time, ten seeds were randomly sampled and mixed with 2 mL of sterile water

Fungicide codification	Active ingredient (ai)	Concentration of the ai (g L^{-1} or g kg^{-1})	Dose[a]
Tm	Thiophanate-methyl	500	100[b]
TmF	Thiophanate-methyl + fluazinam	350 + 52.5	180[b]
C	Carbendazim	500	100[b]
F	Fluazinam	500	200[c]
FldMM	Fludioxonil + metalaxyl-M	25 + 10	100[b]
FiTmPy	Fipronil + t. methyl + pyraclostrobin	250 + 225 + 25	200[b]
Pro	Procymidone	500	200[c]

[a]*mL of commercial product 100 kg^{-1} of seeds*
[b]*Dose recommended for seed treatment*
[c]*Experimental dose*

Table 3.
Chemical fungicides used in the compatibility test.

Treatment	Active ingredient (ai)	Application timing[e]				Dose (L/kg ai ha^{-1})
		1st	2nd	3rd	4th	
Control	—	—	—	—	—	—
SF04	*T. asperellum*[a]	V_4	V_6	—	—	2.0×10^9
IBLF006	*T. harzianum*[b]	V_4	V_6	—	—	2.0×10^8
ESALQ-1306	*T. harzianum*[c]	V_4	V_6	—	—	2.0×10^9
Tricho	*Trichoderma* spp.[d]	V_4	V_6	—	—	2.0×10^6
ESALQ-1306 + Tm	*T. harzianum*[c] +thiophanate-methyl	V_4	V_6			2.0×10^9
				R_1	R_2	0.5
Tm	Thiophanate-methyl	—	—	R_1	R_2	0.5
F	Fluazinam	—	—	R_1	R_2	0.5

[a]*Strain SF04*
[b]*Strain IBLF006*
[c]*Strain ESALQ-1306*
[d]*Not specified on product's label*
[e]*According to soybean phenological stage proposed by [56]*

Table 4.
Description of active ingredients/microorganisms, application timing, and doses of products in the control of Sclerotinia stem rot in soybean crop.

for 1 min. Aliquots of 100 μL were plated in PDA acidified medium (1 mL lactic acid L^{-1} medium) and incubated at 25°C for 20 h. Subsequently, spore germination of *Trichoderma* strains was inhibited with lactophenol-blue cotton solution. Coverslips were placed onto two regions of each plate and 100 spores counted at each one of them under light microscope (Olympus CX40). The result was expressed as the percentage of germinated spores. Despite the pressing need to reduce the use of pesticides in food production systems, extensive monocropping areas are still chemical dependents. In this scenario, effectiveness of biological control agents requires knowledge on their compatibility with the most common pesticides used in crops. Specifically, we accessed the viability of four different *Trichoderma*-based products toward active ingredients (ai) recommended to the control of SSR on soybean crop. First, we performed an in vitro baseline sensitivity study to monitor the inhibition of mycelium growth and proceed with a seed treatment assay of successive application of chemical and biological products to determine killing effect on antagonist's spores.

We used diameter data measured from colonies during a 7-day incubation period to calculate the growth speed index (GSI) of the *Trichoderma* strains in culture medium containing different concentrations of chemical fungicides. Significant effects ($p < 0.001$, CV = 14.55%) were found for the triple interaction of fungicides, concentrations, and bioproducts. Among tested fungicides, thiophanate-methyl (Tm) was the only one with a logistic behavior (all the others adjusted to exponential regressions), requiring doses above 1.36 ppm to affect growth speed of the strains by 50% (**Table 5**). Coefficients of exponential equation of ai carbendazim (C) indicate greater impact of this fungicide on mycelial growth of *Trichoderma* spp., as observed for lower values of coefficient a and higher values of coefficient b compared to all bioproducts. Compatibility responses varied among bioproducts, fungicides, and concentrations, following a trend of the higher the concentration, the lower the development of the antagonist. In general, the fungicide carbendazim (C) was less compatible considering all tested concentrations, opposing to fungicides thiophanate-methyl (Tm) and procymidone (Pro) (**Table 5**). However, procymidone reduced the GSI from 13.6 to 83%, demanding caution in the recommendation of its use associated with bioproducts in foliar sprays, application in the soil, or seed treatment.

To check the compatibility between chemical fungicides and bioproducts in seed treatment, a common agricultural practice, we evaluated the viability of the spores of the antagonists after exposure to the active ingredients. Like the plating assay, in the soybean seed treatment, significant effects ($p < 0.001$, CV = 15.69%) were observed for the triple interaction (fungicides × bioproducts × exposure time). Excluding the association of thiophanate-methyl (Tm) with IBLF006 for a 3-h incubation, all fungicides reduced the germination potential of the spores (**Figure 3**). In general, fungicides that led to smaller reductions in germination were thiophanate-methyl (Tm), procymidone (Pro), and fipronil + t.-methyl + pyraclostrobin (FiTmPy) (germination rates of 87.3, 64.5, and 63.8%, respectively). These results indicate potentiality in their use combined to the bioproducts in the treatment of soybean seeds. The ai fluazinam (alone or associated to thiophanate-methyl) and carbendazim (C) drastically reduced viability of *Trichoderma* strains (up to complete inhibition of spore germination). In vitro assays report insensitivity of *T. harzianum*, *T. stromaticum*, and *T. atroviride* to procymidone [59–61] suggesting its simultaneous use with *Trichoderma* species for the control of *S. sclerotiorum* in tomato [62] and lettuce [63]. Other research demonstrate that thiophanate-methyl is compatible to many strains of *Trichoderma* spp., whereas carbendazim and fluazinam are extremely toxic [64–65], supporting our findings. The insensitiveness to dicarboximide fungicides (such as procymidone) may probably be due to higher transcriptional levels of histidine kinase genes [28]. Interestingly, the *Trichoderma* strains we

Fungicide	SF04	IBLF006	ESALQ-1306	Tricho	SF04	IBLF006	ESALQ-1306	Tricho
	0.1 ppm				1 ppm			
Tm	12.0 Ab	12.6 Aa	12.6 Aa	13.0 Aa	6.7 Bc	12.0 Aa	10.1 Ab	12.2 Aa
TmF	12.1 Aa	8.7 Db	5.9 Dd	7.9 Dc	5.3 Ca	4.0 Db	3.6 Db	3.7 Db
C	1.3 Da	1.2 Fa	1.1 Ga	1.7 Ga	0.4 Ea	1.0 Ga	0.5 Fa	1.0 Fa
F	4.6 Ca	3.0 Eb	4.4 Ea	4.8 Ea	4.7 Ca	3.3 Eb	3.8 Db	3.9 Db
FldMM	1.9 Db	2.0 Gb	2.3 Fb	3.9 Fa0	1.4 Db	2.3 Fa	2.5 Ea	2.7 Ea
FiTmPy	8.6 Bb	9.7 Ca	7.3 Cc	8.8 Cb	5.4 Ca	6.1 Ca	4.7 Cb	5.6 Ca
Pro	8.0 Bb	10.8 Ba	11.2 Ba	11.5 Ba	12.2 Aa	9.3 Bb	9.2 Bb	9.7 Bb
	10 ppm				100 ppm			
Tm	0.3 Cb	1.5 Ba	1.7 Ca	0.9 Bb	0.1 Ca	0.1 Da	0.1 Da	0.2 Da
TmF	0.8 Ca	0.3 Ca	0.5 Da	0.2 Ba	0.1 Ca	0.0 Da	0.1 Da	0.1 Da
C	0.0 Ca	0.3 Ca	0.2 Da	0.2 Ba	0.1 Ca	0.5 Da	0.2 Da	0.2 Da
F	4.6 Aa	2.1 Ab	2.7 Bb	2.6 Ab	3.6 Aa	1.5 Bc	2.3 Bb	2.3 Bb
FldMM	1.8 Ba	2.3 Aa	1.9 Ca	1.9 Aa	1.1 Ba	1.1 Ca	1.0 Ca	1.1 Ca
FiTmPy	0.4 Ca	0.4 Ca	0.3 Da	0.3 Ba	0.4 Ca	0.1 Da	0.2 Da	0.1 Da
Pro	2.1 Bb	2.2 Ab	3.5 Aa	2.0 Ab	3.7 Aa	3.0 Aa	3.6 Aa	3.2 Aa
	1000 ppm				0 ppm			
Tm	0.0 Ca	0.0 Ba	0.0 Ba	0.0 Ba	14.12	12.81	13.61	11.96
TmF	0.0 Ca	0.0 Ba	0.0 Ba	0.0 Ba				
C	0.0 Ca	0.3 Ba	0.1 Ba	0.0 Ba				
F	2.1 Ba	0.5 Ba	0.8 Ba	0.7 Ba				
FldMM	0.3 Ca	0.5 Ba	0.5 Ba	0.3 Ba				
FiTmPy	0.0 Ca	0.0 Ba	0.1 Ba	0.0 Ba				
Pro	3.9 Aa	2.4 Aa	3.5 Aa	3.2 Aa				

Colonies were plated in PDA medium supplemented with chemical fungicides at concentrations ranging from 0 to 1000 ppm
Averages followed by the same uppercase letters, in columns, and lowercase letters, in lines, do not differ significantly by the Tukey test ($p \leq 0.05$)

Table 5.
Growth speed index (GSI) calculated based on Trichoderma mycelium diameter measured daily.

studied showed completely divergent behavior regarding the benzimidazole fungicides: compatible to thiophanate-methyl but sensitive to carbendazim. Differences on tubulin-binding site or degradation/detoxification of the active ingredients could explain selectivity (however, these aspects are yet to be demonstrated).

The simultaneous use of biocontrol agents and pesticides in disease management may allow reduction of recommended doses of chemicals [66]. This possibility could mitigate compatibility problems, as the fungicides applied in low concentrations did not visibly affect *Trichoderma* growth and viability. On the other hand, such phytosanitary strategy often results in synergistic or additive effects in the control of soil-borne diseases. In seed treatment, it is recommended that sowing be done as soon as possible, since spore germination tends to decrease as exposure to chemicals is prolonged. In vitro studies have the advantage of exposing the microorganism as much as possible to the action of the chemical, a fact that does not occur in field conditions where many factors constitute obstacles to this exposure, thus protecting the biological agent. Therefore, there is a possibility of good selectivity of thiophanate-methyl and procymidone under field conditions considering results obtained in the laboratory.

2.7 Sanity of soybean seeds treated with *Trichoderma* spp.

The efficacy of the antagonist in the control of seed pathogens was verified through the blotter test. Four hundred soybean seeds cv. NK7074RR, artificially inoculated or not with *S. sclerotiorum*, were divided in 16 pseudo-samples and distributed in transparent acrylic boxes filled with three sterile sheets of filter paper moistened with distilled water. Boxes were previously disinfected with 70% ($v\ v^{-1}$) ethanol and 2% ($w\ v^{-1}$) sodium hypochlorite solution. Before test installation, soybean seeds were subjected to −20°C for 24 h to retard germination. Artificial inoculation with *S. sclerotiorum* consisted of placing the seeds in Petri dishes completely colonized by the pathogen. Seeds were kept in contact with the fungus until visual detection of the white mycelium on them. Later, the seeds were treated with the bioproducts at the commercial dose recommended by the manufactures and incubated during 1 week at 20 ± 2°C under 12 h photoperiod and then at 12 ± 2°C for equal period with additional wetting of the paper substrate [67]. Seeds were individually checked using a stereomicroscope at the resolution of 30–80× for the identification of typical fungi-fruiting bodies. Using a light microscope, we confirmed the identity of the fungi at species level when necessary.

Seeds can be source of inoculum introducing pathogens to new cultivation areas and increasing diseases in the field. Besides, physiological seed quality can be compromised by deteriorating action of fungi during storage. Alternatively, chemical seed treatment microbiolization ensures seed health by using living microorganisms. In the sanity test, soybean seeds were treated with the bioproducts. We observed predominance, but not exclusivity, of *Fusarium semitectum* (up to 31% of contaminated seeds), *Nigrospora* spp. (12%), and *Aspergillus* spp. (5%). There was also incidence of other fungi such as *Colletotrichum dematium* var. *truncata*, *Phomopsis phaseoli*, *Penicillium* spp., *Periconia* spp., *Rhizopus stolonifer*, and *Cladosporium* spp. (data not shown). All bioproducts reduced the incidence of *Penicillium* spp., *Aspergillus* spp., *R. stolonifer*, *P. phaseoli*, *Cladosporium* spp., and *Periconia* spp. compared to the control. Regarding *Phomopsis* spp., its occurrence is closely dependent on environmental conditions during seed formation and maturation [68], and its presence in the seed lot makes it unfeasible for use. Microbiolization efficacy, examined to all *Trichoderma* strains tested, can ensure sowing of seed lots obtained from fields in which maturation and harvesting coincided with high temperatures and high relative air humidity. Besides, these results suggest not only the known use of *Trichoderma* spp. in the protection of seeds against soil-borne pathogens [37] but against storage fungi, recommending its application at post-harvest. It is likely that microbiolization will promote disease control and favor seed germination, seedling emergence, and seedling early development in the field as it protects the seeds from fungi attack during storage. Moreover, strains IBLF006 and ESALQ-1306 were efficient in reducing incidence of *F. semitectum* (50 and 100%, respectively).

Now considering the soybean seeds artificially inoculated, *S. sclerotiorum* suppressed the development of the microorganisms, as well as the colonization capability of Tricho and IBLF006 strains. On the other hand, ESALQ-1306 showed higher competitiveness against *S. sclerotiorum* reducing its occurrence by 40%. This effect, combined with that found in the soil parasitism test, reiterates the potential of the bioproduct ESALQ-1306 in the containment of the dissemination of SSR through seeds. Taking into consideration soybean seeds may be contaminated by either sclerotia or *S. sclerotiorum* mycelium, the strain ESALQ-1306 was effective in the control of both contamination forms.

Seed microbiolization represents a useful and promising method for the control of seed pathogens (infecting or contaminating the seed lot) and of soil-borne pathogens (as *Fusarium* spp.). It is considered an important method of application of biocontrol agents since it requires a small amount of biological material compared to the quantity needed for soil application. Furthermore, it is an efficient and low-environmental-impact strategy compared to chemical fungicides and may increase the concentration of the antagonist in the soil in medium/long terms turning it suppressive to various pathogens. *Trichoderma*-based products are commercialized as biofungicides, biostimulants, biofertilizers, and even plant growth promoters [37]. We confirmed the in vitro antagonistic activity of the bioproducts against *S. sclerotiorum* and their potential enhancement of seed health and proceeded to test the hypothesis of their effect to promote initial development of soybean seedlings. Seed treatment with the bioproducts did not favor germination though neither impaired the development of normal plantlets. Upon seed challenge with *S. sclerotiorum*, the occurrence of abnormalities varied depending on the bioproduct applied: IBLF006 presented the lowest (3.5%) and SF04 the highest incidence (22%) of abnormal seedlings. Pathogen inoculation reduced germination rate by 8.4%. Despite the greater occurrence of abnormal seedlings, the SF04 strain was detected colonizing the soybean seedlings, as did the strain ESALQ-1306, suggesting competitiveness against *S. sclerotiorum*. In the absence of the phytopathogen, the treatments did not differ among them, with average colonization of developed seedlings of 60.7%.

All bioproducts accelerated emergence speed index on sand seedbed test. The index practically doubled compared to the untreated control. This result indicates improvement in the physiological quality of soybean seeds inoculated with *Trichoderma* spp. On the other hand, the faster soybean seedlings develop, the less the seeds will be exposed to soil-borne pests. The bioproducts did not change the analyzed biometric variables, except for root length and shoot fresh weight. Under *S. sclerotiorum* infection, seedlings showed higher root growth when treated with strain ESALQ-1306 (8.26 cm), followed by SF04 and Tricho. Strain IBLF006 was like the control (6.83 cm). ESALQ-1306 promoted root growth also in uninfected seeds (8.30 cm). *Trichoderma* spp. treatment increased in up to 36.6% aerial fresh weight of artificially inoculated soybean (equivalent to 8.96 g) and reduced SSR incidence in 2.5 folds (strains ESALQ-1306 and SF04). Beneficial effects of *Trichoderma* spp. on plant development are reported in rice [76], melon [69], sorghum [70], and wheat [71, 72]. In soybean crop, growth promotion is related to the synthesis of 1-aminocyclopropane-1-carboxylate deaminase, indole acetic acid (IAA), siderophores, and biological nitrogen fixation [73, 74]. The greater root development found in our study is possibly associated with the synthesis of IAA and reduction of ethylene levels in plants. Consequently, plantlets uptake more nutrients [75] including iron and nitrogen and increase aboveground vegetative growth. This auxin-ethylene cross talk induces mitogen-activated protein kinases and proteins involved in carbohydrate metabolism and photosynthesis [77]. Likewise, signaling of growth regulators may have favored the rapid emergence of seedlings, a fundamental feature for successful crop establishment especially facing biotic and abiotic stresses.

2.8 Effect of *Trichoderma* spp. on soybean germination and early development and effect of *Trichoderma* spp. on physiological quality of soybean seeds

The standardization of the seed germination and seedling emergence on sand seedbed tests [78] to access possible effects of the antagonist on physiological quality of soybean seeds is an important procedure. Germination test consisted of four replicates of 50 seeds each placed in filter paper rolls as recommended [78]. Soybean seeds were artificially inoculated with *S. sclerotiorum* as described (item 2.7), treated with the bioproducts, and further incubated in a growth chamber under 12 h photoperiod and 25 ± 2°C. Eight days post-incubation, we evaluated the number of normal, abnormal, and infected plantlets with SSR or *Trichoderma* spp. Fungi presence was confirmed with a stereomicroscope. We considered abnormal those plantlets without the main root or displaying hypocotyl abnormalities, such as its absence, damages, or rotting [79]. Seedling emergence test was carried out in a greenhouse. Soybean seeds were placed in plastic trays filled with sterile sand. The substrate was uniformly moistened according to the calculus of its water retention capacity [78]. The number of emerged plantlets was checked daily since the day of the first normal seedling emergence. Counting continued until the 13th day. The emergence speed index was calculated according to Maguire [80]:

$$ESI = E_1/N_1 + E_2/N_2 + \ldots + E_n/N_n \tag{1}$$

where ESI = emergence speed index; E_1, E_2, ... E_n = number of normal seedlings obtained at the first, second, and at the nth counting; and N_1, N_2, ... N_n = number of days from sowing to the first, second, and nth counting.

The register of the number of plantlets with abnormalities, with necrotic cotyledons, and infected with SSR, as well as shoot and root lengths (cm) and fresh and dry weights (g), is very important in this case or evaluation. Standard germination test followed a randomized design with four replications of 50 seeds each, whereas the seedling emergence on sand test was carried out in a completely randomized block design with four replicates of 200 seeds each. The treatments consisted of the four biological products and a control (without the antagonist) inoculated or not with *S. sclerotiorum*.

2.9 In vivo biocontrol of *S. sclerotiorum* and biological control of soybean SSR under field conditions

After laboratory and greenhouse experiments, we conducted a field study at a commercial soybean crop geo-referenced at 19°12′54″S and 47°56′58″W, 947 m of altitude, during the summer season (from December/2009 to April/2010). Climatological data [maximum and minimum temperatures (°C), relative air humidity (%), and pluvial precipitation (mm)] were obtained from the weather station located at the farm. Soil was classified as a ferralsol, and the field had previous report of SSR occurrence. Sowing was done with 15 seeds per linear meter using soybean cultivar BRS Valiosa RR (susceptible to SSR) at a final stand of 10 plants m^{-1}. Crop conduction was according to Embrapa [81]. Experimental design was in random blocks with seven treatments and a control (**Table 3**), with four replications. Each plot consisted of six rows of 5 m length and 0.5 m apart totalizing an area of 480 m^2. The four central rows despising 0.5 m from both edges were considered as the useful plot. Spraying was done with a CO_2 pressurized costal sprayer equipped with XR110.02 nozzles at a volume of 200 L ha^{-1}. Environmental conditions were constantly monitored during application of the (bio)products ranging from 27.2 to 34.3°C, 47 to 65% of relative air humidity, and winds of 0 to 5 km h^{-1}. The titer of the *Trichoderma*-based products was calibrated by the viability test in PDA medium (viable conidia mL^{-1}) after incubation at 25 ± 2°C for 5 days. SSR incidence and severity on soybean plants were evaluated at phenological stages R_4, $R_{5.2}$, and $R_{5.5}$ [56]. The data were used to calculate the disease index (% incidence

× % severity) [82] and the area under the disease progress curve (AUDPC) [83]. The severity of SSR was estimated by a visual scale [82] assigning percentages of 5–90% of the symptoms of the disease in individualized soybean plants. Manual harvesting of the useful plots was carried out at the R_8 stage. The productivity was obtained after mechanical track of pods, followed by correction of the moisture content of grains to 12% and extrapolation of the data to kg ha^{-1}. After harvest, the sclerotia were separated from the grains with the aid of sieves and their weight (g) determined per hectare.

Trichoderma species are potential biocontrol agents against a range of plant fungal pathogens. Some examples include *R. solani* [84], *F. oxysporum* f. sp. *melonis* [69], *Puccinia triticina* [85], *C. sublineolum* [70], and *P. graminis* [85]. Though plant-pathogen-*Trichoderma* interactions have been extensively studied even at the molecular level [86–88], most of the research are conducted under laboratory or greenhouse conditions. To evaluate the suppressive effect of *Trichoderma*-based products on SSR providing a means of controlling the disease, we applied the bioproducts in a commercial soybean field. The efficacy of the biological products was compared to chemical fungicides frequently used in soybean crop. Chemicals were evaluated alone or associated to *Trichoderma* spp. AUDPC values were lower in treatments with the fungicide fluazinam and with thiophanate-methyl applied in sequence to the use of strain ESALQ-1306 (**Table 6**). A possible synergistic effect is suggested upon association of biological and chemical products once their spraying alone led to averages statistically equal to the control (without spraying). In this association, application of the antagonist first weakens the pathogen, and the fungicide kills it. This strategy allows rotation of active ingredients with different mechanisms of action in the field with potential reduction of resistant populations of *S. sclerotiorum*. The low control efficacy observed in the application of the chemical fungicide thiophanate-methyl alone requires attention to possible loss of sensitivity of the pathogen. A resistant *S. sclerotiorum* strain was first reported in common bean crop in Brazil in 2015 [23]. Resistance was conferred by a point mutation of a leucine by a phenylalanine at position 240 of the β-tubulin gene.

The use of *Trichoderma* spp. as exclusive control method was not enough to reduce the severity of SSR (**Table 6**). Regarding soybean grain yield, however, strains ESALQ-1306 and SF04 maintained productivity at high levels and increased

Treatments	AUDPC	INCID. (%)	AUDI	SCLE. (g)	TGW (g)	YIELD (kg ha^{-1})
Control	909.4 B	25.5 B	20130.0 B	6.6 A	127.6 B	1942.5 C
SF04	685.0 B	9.2 A	5296.9 A	2.4 A	147.2 A	2890.0 A
IBLF006	766.9 B	10.8 A	7319.4 A	5.8 A	136.4 B	2523.3 B
ESALQ-1306	651.2 B	9.8 A	5005.0 A	2.2 A	145.8 A	2897.5 A
Tricho	656.9 B	12.2 A	5877.5 A	4.6 A	144.5 A	2452.5 B
ESALQ-1306 + Tm	532.5 A	10.5 A	4258.8 A	2.5 A	146.1 A	2765.0 A
Tm	658.1 B	12.8 A	6690.6 A	3.4 A	143.3 A	2945.0 A
F	440.6 A	7.5 A	2620.6 A	1.9 A	151.5 A	3015.0 A

AUDPC, area under the disease progress curve; INCID, disease incidence; AUDI, area under the disease index (% incidence × % severity); SCLE, sclerotia weight; TGW, a thousand grain weight averages followed by different uppercase letters, in columns, are statistically different by the Scott-Knott test ($p \leq 0.05$).

Table 6.
Control of Sclerotinia stem rot in soybean crop due to foliar spraying of Trichoderma-based products and chemical fungicides.

by 1.5-fold over that of the untreated control. The same result was not accomplished by Tricho or IBLF006, which presented intermediate yields, partially justified by inconsistencies in spraying due to their formulation (WP). The mixture of solid in aqueous medium is not stable and demands constant agitation to remain homogeneous, which may have been compromised by the application equipment we adopted. Biocontrol agents and chemical fungicides or their association were not able to completely prevent recurrence of the disease in the field, since they did not completely reduce the formation of sclerotia in only one year of field management. Results from our in vitro assays, though, suggest that the sclerotia were not viable in the plots treated with the bioproducts. This assumption may be supported by sclerotia weight which was one- to threefold lower compared to untreated control. Those values represent 1.9–4.9 kg of sclerotia ha^{-1} contrasting to 5.5 kg ha^{-1} observed in the control.

Disease symptoms were not attenuated in plants treated with the bioproducts; however, they showed significant reduction in SSR incidence and lower disease index. Incidence is the most important parameter when it comes to SSR field evaluation. Disease index estimates the damage to the plant both by the number of diseased plants (incidence) and by the lesion length (severity) [82]. Application of bioproducts alone reduced the index by 64–75%. It is important to mention that biological control does not promote total eradication of phytopathogens but the maintenance of the population at levels enough not to cause economic damages to the crop. In our study, this was reflected by the productivity increase of up to 35 bushels ha^{-1} in relation to the untreated control, an income of US$297 ha^{-1}.

3. Conclusions

In conclusion, we report the use of *Trichoderma* as a soybean seedling growth promoter and as a biological control agent of SSR acting synergistically to thiophanate-methyl. We found strains with in vitro hyperparasitic activity and capable of killing sclerotia. After foliar application on soybean crop, *Trichoderma* strains start soil parasitism preventing ejection of ascocarps and ascospores, as observed in vitro through sclerotia germination and colonization, leading to reduced disease incidence under field conditions. As a result, higher grain yield is achieved. Chemical fungicides may be used simultaneously with bioproducts upon dose reduction (this aspect is yet to be demonstrated). Formulation should be approached with caution during bioproducts' development as it interferes in viability and efficacy. For example, the strain ESALQ-130 showed in vitro antagonism similar to that of other strains tested. However, its formulation (CS) may have favored seeds and field applications. Despite problems on technology application of WP, suspensions are not always stable during storage, as particles may sediment and form a two-phase system that no longer resuspend. Many environmental factors interact with *Trichoderma*-based products affecting their efficacy under field conditions. We encourage that laboratory and greenhouse studies proceed to the field to confirm disease control. Quality of biological products is a threshold to efficacy and must be constantly monitored. There are still gaps in obtaining new formulations, selecting potent strains, evaluating adequate application technologies, and accessing *Trichoderma* spp. performance in cold soils (temperatures lower than 15°C). Further research should be conducted aiming at improving *Trichoderma* recommendation in agriculture, regarding dosage for different crops, intervals, and application timing.

Acknowledgements

The authors would like to thank Coordenação de Aperfeiçoamento de Pessoal de Nível Superior—CAPES, Conselho Nacional de Desenvolvimento Científico e Tecnológico—CNPq and Fundação de Amparo à Pesquisa do Estado de Minas Gerais, FAPEMIG, for financial support. Special thanks go to Fabio J. Carvalho for the statistical advice provided. The third author acknowledges PPGA-UFU and CAPES for PNPD scholarship.

Conflict of interest

There is no conflict of interest in this paper.

Author details

Fernando C. Juliatti*, Anakely A. Rezende, Breno Cezar Marinho Juliatti and Tâmara P. Morais
Plant Protection Laboratory, Institute of Agricultural Sciences, Federal University of Uberlândia, Uberlândia, MG, Brazil

*Address all correspondence to: juliatti@ufu.br

References

[1] Martin-Guay MO, Paquette A, Dupras J, Rivest D. The new green revolution: Sustainable intensification of agriculture by intercropping. Science of the Total Environment. 2018;**615**:767-772

[2] Pellegrini P, Fernández RJ. Crop intensification, land use, and on-farm energy-use efficiency during the worldwide spread of the green revolution. Proceedings of the National Academy of Sciences of the United States of America. 2018;**115**(10):2335-2340

[3] Grapputo A, Boman S, Lindstrom L, Lyytinen A, Mappes J. The voyage of an invasive species across continents: Genetic diversity of North American and European Colorado potato beetle populations. Molecular Ecology. 2005;**14**(14):4207-4219

[4] Hovmoller MS, Yahyaoui AH, Milus EA, Justesen AF. Rapid global spread of two aggressive strains of a wheat rust fungus. Molecular Ecology. 2008;**17**(17):3818-3826

[5] Amaya OS, Restrepo OD, Argüelles J, Garramuño EA. Evaluación del comportamiento del complejo *Spodoptera* con la introducción de algodón transgénico al Tolima, Colombia. Corpoica Ciencia y Tecnología Agropecuaria. 2009;**10**(1):24-32

[6] Singh RP, Hodson DP, Huerta-Espino J, Jin Y, Bhavani S, Njau P, et al. The emergence of Ug99 races of the stem rust fungus is a threat to world wheat production. Annual Review of Phytopathology. 2011;**49**(1):465-481

[7] Cortez-Mondaca E, Pérez-Márquez J, Sifuentes-Ibarra E, Garcia-Gutierrez C, Valenzuela-Escoboza FA, Camacho-Baez JR. Impacto del cambio climático sobre insectos en Sinaloa; el caso palomillo de la papa Phthorimaea operculella Zeller 1873 (Lepidoptera: Gelechiidae). In: Flores-Campana LM, Moran-Ângulo RE, Karam-Quiniones C, editors. Sinaloa y el Cambio Climático Global. Universidad Autónoma de Sinaloa Instituto de Apoyo a la Investigación e Innovación México. 2014. pp. 219-235

[8] Sun SL, Zhu ZD, Zhang JL, Mei L. Outbreak of *Choanephora* stem rot caused by *Choanephora cucurbitarum* on Quinoa (*Chenopodium quinoa*) in China. Plant Disease. 2018. DOI: 10.1094/PDIS-12-17-1922-PDN [Epub ahead of print]

[9] Carvalho FP. Pesticides, environment, and food safety. Food and Energy Security. 2017;**6**(2):48-60

[10] Juliatti FC, Polloni LC, Morais TP, Zacarias NRS, Silva EA, Juliatti BCM. Sensitivity of *Phakopsora pachyrhizi* populations to dithiocarbamate, chloronitrile, triazole, strobilurin, and carboxamide fungicides. Bioscience Journal. 2017;**33**(4):933-943

[11] Ghorbanpour M, Omidvari M, Abbaszadeh-Dahaji P, Omidvar R, Kariman K. Mechanisms underlying the protective effects of beneficial fungi against plant diseases. Biological Control. 2018;**117**:147-157

[12] Harman GE, Howell CR, Viterbo A, Chet I, Lorito M. Trichoderma species—Opportunistic, avirulent plant symbionts. Nature Reviews. Microbiology. 2004;**2**:43-56

[13] Vos CM, De Cremer K, Cammue BPA, De Coninck B. The toolbox of Trichoderma spp. in the biocontrol of *Botrytis cinerea* disease. Mol. Plant Pathology. 2015;**16**:400-412

[14] Błaszczyk L, Siwulski M, Sobieralski K, Lisiecka J, Jedryczka M. Trichoderma spp., application and prospects for use in organic farming and industry.

Journal of Plant Protection Research. 2014;**54**(4):309-317

[15] Scharf DH, Brakhage AA, Mukherjee PK. Gliotoxin—Bane or boon? Environmental Microbiology. 2016;**18**:1096-1109

[16] Toghueoa RMK, Ekea P, Gonzálezb IZ, de Aldanab BRV, Nanaa LW, Boyoma FF. Biocontrol and growth enhancement potential of two endophytic *Trichoderma* spp. from *Terminalia catappa* against the causative agent of common bean root rot (*Fusarium solani*). Biological Control. 2016;**96**:8-20

[17] Tuão Gava CA, Pinto JM. Biocontrol of melon wilt caused by *Fusarium oxysporum* Schlect *f. sp. melonis* using seed treatment with Trichoderma spp. and liquid compost. Biological Control. 2016;**97**:13-20

[18] Vinale F, Sivasithamparam K, Ghisalberti EL, Marra R, Woo SL, Lorito M. Trichoderma-plant-pathogen interactions. Soil Biology & Biochemistry. 2008;**40**:1-10

[19] Smolińska U, Kowalska B, Kowalczyk W, Szczech M. The use of agro-industrial wastes as carriers of Trichoderma fungi in the parsley cultivation. Scientia Horticulturae. 2014;**179**:1-8

[20] Forster R. Inativação do vírus do mosaico comum do fumo pelo filtrado de culturas de *Trichoderma sp*. Bragantia. 1950;**10**(5):139-148

[21] Bettiol W, Morandi MAB, Pinto ZV, Paula Junior TJ, Corrêa EB, Moura AB, et al. Produtos comerciais à base de agentes de biocontrole de doenças de plantas. Jaguariúna: Embrapa Meio Ambiente; 2012

[22] Meyer MC, Campos HD, Godoy CV, Utiamada CM. Ensaios cooperativos de controle biológico de mofo branco na cultura da soja—safras 2012 a 2015. Londrina: Embrapa Soja; 2016

[23] Lehner MS, Paula Junior TJ, Silva RA, Vieira RF, Carneiro JES, Schnabel G, et al. Fungicide sensitivity of *Sclerotinia sclerotiorum*: A thorough assessment using discriminatory dose, EC50, high-resolution melting analysis, and description of new point mutation associated with thiophanate-methyl resistance. Plant Disease. 2015;**99**(11):1537-1543

[24] Di YL, Zhu ZQ, Lu XM, Zhu FX. Baseline sensitivity and efficacy of trifloxystrobin against *Sclerotinia sclerotiorum*. Crop Protection. 2016;**87**:31-36

[25] Meyer MC, Campos HD, Godoy CV, Utiamada CM, Pimenta CB, Jaccoud Filho DS, et al. Eficiência de fungicidas para controle de mofo-branco (*Sclerotinia sclerotiorum*) em soja, na safra 2016/17: resultados sumarizados dos ensaios cooperativos. Londrina: Embrapa Soja; 2017

[26] Hou YP, Mao XW, Lin SP, Song XS, Duan YB, Wang JX, et al. Activity of a novel succinate dehydrogenase inhibitor fungicide pyraziflumid against *Sclerotinia sclerotiorum*. Pesticide Biochemistry and Physiology. 2018;**145**:22-28

[27] Hu S, Zhang J, Zhang Y, He S, Zhu F. Baseline sensitivity and toxic actions of boscalid against *Sclerotinia sclerotiorum*. Crop Protection. 2018;**110**:83-90

[28] Li JL, Zhu F, Li J. Expression of the histidine kinase gene Sshk correlates with dimethachlone resistance in *Sclerotinia sclerotiorum*. Phytopathology. 2018. DOI: 10.1094/PHYTO-05-18-0156-R [Epub ahead of print]

[29] Mao XW, Li JS, Chen YL, Song XS, Duan YB, Wang JX, et al. Resistance risk assessment for fluazinam in *Sclerotinia sclerotiorum*. Pesticide Biochemistry and Physiology. 2018;**144**:27-35

[30] Kumar S, Thakur M, Rani A. Trichoderma: Mass production, formulation, quality control, delivery and its scope in commercialization in India for the management of plant diseases. African Journal of Agricultural Research. 2014;**9**(53):3838-3852

[31] Garcia RA, Juliatti FC. Avaliação da resistência da soja a *Sclerotinia sclerotiorum* em diferentes estádios fenológicos e períodos de exposição ao inóculo. Tropical Plant Pathology. 2012;**37**(3):196-203

[32] Araújo WL, Maccheroni Junior W, Aguilarvildoso CI, Barroso PAV, Saridakis HO, Azevedo JL. Variability and interactions between endophytic bacteria and fungi isolated from leaf tissues of citrus rootstocks. Canadian Journal of Microbiology. 2001;**47**:229-236

[33] Mello SCM, Ávila ZR, Braúna LM, Pádua RR, Gomes D. Cepas de *Trichoderma* para el control biológico de *Sclerotium rolfsii* Sacc. Fitosanidad. 2007;**11**(1):3-9

[34] Bardin SD, Huang HC. Research on biology and control of *Sclerotinia* diseases in Canada. Canadian Journal of Plant Pathology. 2001;**23**(1):88-98

[35] Bell DK, Wells HD, Markham CR. *In vitro* antagonism of *Trichoderma* species against six fungal plant pathogens. Phytopathology. 1982;**72**(4):379-382

[36] Vale FXR, Fernandes Filho EI, Liberato JR. QUANT: A software plant disease severity assessment. In: International Congress of Plant Pathology, 8. Christchurch: International Society for Plant Pathology; 2003. p. 105

[37] Woo SL, Ruocco M, Vinale F, Nigro M, Marra R, Lombardi N, et al. *Trichoderma*-based products and their widespread use in agriculture. Open Mycology Journal. 2014;**8**:71-126

[38] Milner RJ, Huppatz RJ, Swaris SC. A new method for assessment of germination of *Metarhizium* conidia. Journal of Invertebrate Pathology. 1991;**57**:121-123

[39] Stentelaire C, Antoine N, Cabrol C, Feron G, Durand A. Development of a rapid and highly sensitive biochemical method for the measurement of fungal spore viability: An alternative to the CFU method. Enzyme and Microbial Technology. 2001;**29**:560-566

[40] Delalibera I Jr, Humber RA, Hajek AE. Preservation of *in vitro* cultures of the mite pathogenic fungus *Neozygites tanajoae*. Canadian Journal of Microbiology. 2004;**50**:579-586

[41] Magalhães B, Faria M, Frazão H. A technique to estimate the conidial viability of *Metarhizium flavoviride* Gams & Rozsypal (Hyphomycetes) formulated in vegetable oil. Anais da Sociedade Entomológica do Brasil. 1997;**26**(3):569-572

[42] Pomella AWV, Ribeiro RTS. Controle biológico com *Trichoderma* em grandes culturas—uma visão empresarial. In: Bettiol W, Morandi MAB, editors. Biocontrole de doenças de plantas: Uso e perspectivas. Jaguariúna: Embrapa Meio Ambiente; 2009. pp. 239-244

[43] Keswani C, Bisen K, Singh V, Sarma BK, Singh HB. Formulation technology of biocontrol agents: Present status and future prospects. In: Arora N, Mehnaz S, Balestrini R, editors. Bioformulations: For Sustainable Agriculture. New Delhi: Springer; 2016. pp. 35-52

[44] Mariano RdLR. Métodos de seleção in vitro para o controle biológico de patógenos de plantas. In: Fernandes JMC, Prestes AM, Picinini EC, editors. Revisão Anual de Patologia de Plantas. Passo Fundo, EMBRAPA-Trigo; 1993. pp. 369-409

[45] Louzada GAS, Carvalho DDC, Mello SCM, Lobo M Jr, Martins I, Braúna LM. Antagonist potential of *Trichoderma spp*. from distinct agricultural ecosystems against *Sclerotinia sclerotiorum* and *Fusarium solani*. Biota Neotropica. 2009;**9**(3):145-149

[46] Agrios GN. Plant Pathology. fifth ed. New York: Elsevier Academic Press; 2005

[47] de Melo IS. Potencialidades de utilização de *Trichoderma spp*. no controle biológico de doenças de plantas. In: de Melo IS, editor. Controle Biológico de Doenças de Plantas. Jaguariúna: EMBRAPA-CNPMA; 1991. pp. 135-156

[48] de Melo IS, Costa FG. Desenvolvimento de uma formulação granulada a base de *Trichoderma harzianum* para controle de fitopatógenos. Jaguariúna: Comunicado Técnico; 2005

[49] Carvalho DDC, Oliveira TAS, Braúna LM, Mello SCM. Isolados de *Trichoderma sp.* antagônicos a *Fusarium oxysporum*. Brasília: Comunicado Técnico; 2008

[50] Mendonza JLH, Pérez MIS, Prieto JMG, Velásquez JDQ, Olivares JGG, Langarica HRG. Antibiosis of Trichoderma spp. strains native to northeastern Mexico against the pathogenic fungus Macrophomina phaseolina. Brazilian Journal of Microbiology. 2015;**46**(4):1093-1101

[51] Matei S, Matei GB, Cornea P, Popa G. Characterization of soil Trichoderma isolates for potential biocontrol of plant pathogens. Soil Forming Factors and Processes from the Temperate Zone. 2011;**10**:29-37

[52] Manjunath M, Singh A, Tripathi AN, Prasanna R, Rai AB, Singh B. Bioprospecting the fungicides compatible *Trichoderma asperellum* isolate effective against multiple plant pathogens *in vitro*. Journal of Environmental Biology. 2017;**38**(4):553-560

[53] Görgen CA, Silveira Neto AN, Carneiro LC, Ragagnin V, Lobo Junior M. White mold control with mulch and *Trichoderma harzianum* 1306 on soybean. Pesquisa Agropecuária Brasileira. 2009;**44**(12):1583-1590

[54] Bossola JJ, Russel LD. Electron Microscopy. 2nd ed. Boston: Jones and Bartlett Publishers; 1998

[55] Tanaka OAF, Kitajima EW. Treinamento de Técnicas em Microscopia Eletrônica de Varredura. Piracicaba: ESALQ – NAP/MEPA; 2009

[56] Fehr WR, Caviness CE. Stages of Soybean Development, Yowa State University of Science and Technology. Ames: Cooperative Extension Service; 1977

[57] Lilly VG, Barnett HL. Growth. In: Physiology of the Fungi. New York: Mcgraw-Hill; 1951

[58] Caires AM. Reação de genótipos de soja transgênicos e convencionais à podridão branca da haste [Master's Dissertation]. Uberlândia: Universidade Federal de Uberlândia; 2011

[59] McLean KL, Hunt JS, Stewart A. Compatibility of the biocontrol agent *Trichoderma harzianum* C52 with selected fungicides. New Zealand Plant Protection. 2001;**54**:84-88

[60] McLean KL, Hunt JS, Stewart A, Wite D, Porter IJ, Villalta O. Compatibility of a *Trichoderma atroviride* biocontrol agent with management practices of *Allium* crops. Crop Protection. 2012;**33**:94-100

[61] Paula Júnior TJ, Vieira RF, Rocha PRR, Bernardes A, Costa EL, Carneiro JES, et al. White mold intensity on common bean in response to plant

density, irrigation frequency, grass mulching, *Trichoderma* spp., and fungicide. Summa Phytopathologica. 2009;**35**(1):44-48

[62] Aguiar RA, Cunha MG, Lobo Junior M. Management of white mold in processing tomatoes by Trichoderma spp. and chemical fungicides applied by drip irrigation. Biological Control. 2014;**74**:1-5

[63] Silva MAF, Moura KE, Moura KE, Salomão D, Patricio FRA. Compatibility of *Trichoderma* isolates with pesticides used in lettuce crop. Summa Phytopathologica. 2018;**44**(2):137-142

[64] Ribas PP, Paz ICP, Santin RCM, Guimarães AMG, Silva ME, Matsumura ATS. Toxicidade de princípios ativos de fungicidas comerciais sobre *Trichoderma* spp. In: Congresso Brasileiro de Fitipatologia, 42; Rio de Janeiro, CD-ROM. 2009

[65] Lobo M Jr, Geraldine AM, Carvalho DDC. Controle biológico de patógenos habitantes do solo com *Trichoderma* spp., na cultura do feijoeiro comum. Santo Antônio de Goiás: Circular Técnico; 2009

[66] Kumar R, Singh SK, Yadav S, Kumar R, Choubey AK, Kumari A. Compatibility of *Trichoderma viride* with different fungicide and organic cake. Journal of Pharmacognosy and Phytochemistry. 2018;**7**(2):2398-2401

[67] Brasil. Ministério da Agricultura, Pecuária e Abastecimento. Manual de análise sanitária de sementes. Brasília: Mapa/ACS; 2009

[68] Martins CAO, Sediyama CS, Moreira MA, Rocha VS, Oliveira MGA, Gomes JLL. Qualidade sanitária das sementes de soja sem as três lipoxigenases. Revista Ceres. 2003;**50**(288):241-249

[69] Martínez-Medina A, Alguacil MDM, Pascual JA, Van Wees SCM. Phytohormone profiles induced by *Trichoderma* isolates correspond with their biocontrol and plant growth-promoting activity on melon plants. Journal of Chemical Ecology. 2014;**40**:804-815

[70] Saber WIA, Ghoneem KM, Rashad YM, Al-Askar AA. *Trichoderma harzianum* WKY1: An indole acetic acid producer for growth improvement and anthracnose disease control in sorghum. Biocontrol Science and Technology. 2017;**27**:654-676

[71] Bernat P, Nykiel-Szymańska J, Gajewska E, Różalska S, Stolarek P, Dackowa J, et al. *Trichoderma harzianum* diminished oxidative stress caused by 2,4-dichlorophenoxyacetic acid (2,4-D) in wheat, with insights from lipidomics. Journal of Plant Physiology. 2018;**229**:158-163

[72] El-Sharkawy HHA, Rashad YM, Ibrahim SA. Biocontrol of stem rust disease of wheat using arbuscular mycorrhizal fungi and *Trichoderma* spp. Physiological and Molecular Plant Pathology. 2018;**103**:84-91

[73] Zhang F, Ge H, Zhang F, Guo N, Wang Y, Chen L, et al. Biocontrol potential of *Trichoderma harzianum* isolate T-aloe against *Sclerotinia sclerotiorum* in soybean. Plant Physiology and Biochemistry. 2016;**100**:64e74

[74] Zhang F, Chen C, Zhang F, Gao L, Liu J, Chen L, et al. *Trichoderma harzianum* containing 1-aminocyclopropane-1-carboxylate deaminase and chitinase improved growth and diminished adverse effect caused by *Fusarium oxysporum* in soybean. Journal of Plant Physiology. 2017;**210**:84-94

[75] Zhang W. Study on the Integrated Control of Soybean *Sclerotinia* Stem Rot Caused by *Sclerotinia sclerotiorum*. Harbin: Northeast Agricultural University; 2009

[76] Abdel-Fattah GM, Shabana YM, Ismail AE, Rashad YM. *Trichoderma harzianum*: A biocontrol agent against *Bipolaris oryzae*. Mycopathologia. 2007;**164**:81-89

[77] López-Bucio J, Pelagio-Flores R, Herrera-Estrella A. *Trichoderma* as biostimulant: Exploiting the multilevel properties of a plant beneficial fungus. Scientia Horticulturae. 2015;**196**(30):109-123

[78] Brasil. Ministério da Agricultura, Pecuária e Abastecimento. Regras para análise de sementes. Brasília: Mapa/ACS; 2009

[79] Nakagawa J. Testes de vigor baseados no desempenho das plântulas. In: Krzyzanowski FC, Vieira RD, França Neto JB, editors. Vigor de sementes: conceitos e testes. Londrina: ABRATES; 1999

[80] Maguire JD. Speed of germination-aid selection and evaluation for seedling emergence and vigor. Crop Science. 1962;**2**(1):176-177

[81] Embrapa Soja. Tecnologias de Produção de soja-região central do Brasil—2014. Embrapa Soja, Londrina: Sistemas de Produção; 2013

[82] Juliatti FC, Juliatti FC. Podridão branca da haste da soja: Manejo e uso de fungicidas em busca da sustentabilidade nos sistemas de produção. Uberlândia: Composer; 2010

[83] Campbell CL, Madden LV. Introduction to Plant Disease Epidemiology. New York: John Wiley & Sons; 1990

[84] Asad SA, Ali N, Hameed A, Khan SA, Ahmad R, Bilal M, et al. Biocontrol efficacy of different isolates of *Trichoderma* against soil borne pathogen *Rhizoctonia solani*. Polish Journal of Microbiology. 2014;**63**:95-103

[85] El-Sharkawy HHA, Tohamey S, Khalil AA. Combined effects of *Streptomyces viridosporus* and *Trichoderma harzianum* on controlling wheat leaf rust caused by *Puccinia triticina*. Plant Pathology Journal. 2015;**14**:182-188

[86] Atanasova L, Le Crom S, Gruber S, Coulpier F, Seidl-Seiboth V, Kubicek CP, et al. Comparative transcriptomics reveals different strategies of *Trichoderma* mycoparasitism. BMC Genomics. 2013;**14**:121

[87] Zhang X, Harvey PR, Stummer BE, Warren RA, Zhang G, Guo K, et al. Antibiosis functions during interactions of *Trichoderma afroharzianum* and *Trichoderma gamsii* with plant pathogenic *Rhizoctonia* and *Pythium*. Functional & Integrative Genomics. 2015;**15**(5):599-610

[88] Sharma V, Salwan R, Sharma PN, Kanwar SS. Elucidation of biocontrol mechanisms of *Trichoderma harzianum* against different plant fungal pathogens: Universal yet host specific response. International Journal of Biological Macromolecules. 2017;**95**:72-79

Chapter 4

A Review Study on the Postharvest Decay Control of Fruit by *Trichoderma*

Ramsés González-Estrada, Francisco Blancas-Benítez, Beatriz Montaño-Leyva, Cristina Moreno-Hernández, Luz Del Carmen Romero-Islas, Jovita Romero-Islas, Rosa Avila-Peña, Anelsy Ramos-Guerrero, Angel Fonseca-Cantabrana and Porfirio Gutierrez-Martinez

Abstract

This chapter consists of an overview with the most relevant results about the efficacy of *Trichoderma* on postharvest disease control. The results of investigations demonstrate that this fungus can control several phytopathogens in different fruits. Postharvest losses represent a major problem in several countries. The constant application of fungicides not only at field but also at postharvest stage has led to microbial resistance cases, which make the control of these pathogens difficult. Biological control is a promising alternative to chemical fungicide applications. In this sense, an eco-friendly alternative and effective approach for controlling diseases is the use of microbial antagonists like *Trichoderma*, which have several mechanisms of action to stop disease development. A crucial treat in biological control is related to the maintenance of microbial viability and efficacy, that is why other technologies like their incorporation into edible films and coatings, nanotechnology, microbial mixtures, among others have been applied in combination with *Trichoderma* successfully. An enhancement in biocontrol activity is achieved when alternative systems are combined like GRAS substances, biopolymers, and other antagonists. Thus, *Trichoderma* is an eco-friendly alternative to threat postharvest diseases as an alternative to chemical treatments.

Keywords: *Trichoderma*, postharvest, pathogens, fruits, alternative systems

1. Introduction

Postharvest diseases represent a major problem through the world causing significant losses at postharvest stage [1]. Postharvest treatments play an important role in the quality preservation of commodities; however, in developed countries, the inadequate storage and transportation systems favor the establishment of diverse pathogens [2]. Traditionally, postharvest disease management is carried out by the application of chemical fungicides; however, environmental and health issues as well as microbial resistance play an important role in the development of new strategies for controlling diseases [3]. Biological control is an eco-friendly alternative for postharvest disease control. Antagonists can be isolated from diverse sources like soil, fruits, leaves, and from extreme conditions as marine environments [2, 4]. *Trichoderma* is recognized due to their effectiveness in controlling several pathogens in diverse fruits like strawberry (*Botrytis cinerea*), citrus (*Penicillium italicum*), kiwifruit (*Botrytis cinerea*), banana (*Colletotrichum musae*), guava (*Rhizopus* spp.), among others [2]. The efficacy of *Trichoderma* is related to several mechanism of action reported like competition, antibiosis, parasitism (involving lytic enzymes), and the induction of plant defenses [5–7]. Biocontrol activity of *Trichoderma* can be enhanced by the combination of this antagonist with other control systems like the use of GRAS substances, encapsulation in polymeric matrices (chitosan), physical methods, etc. The aim of this chapter was to summarize information about the efficacy and application of *Trichoderma* alone or in combination with other alternative methods in different fruits against the most important postharvest pathogens.

2. Fruits: importance at international level

In recent years, the demand for food has increased dramatically, while the world population has increased by 70%, food consumption per capita has increased more than 20%. According to some reports, it is projected that the production of crops for the year 2030 is 70% higher than the productions that are currently available [8]. Fruits and vegetables will play an important role by providing essential nutrients for the population's diet, both in developed and developing countries, in addition to being associated with the reduction of the risk of suffering from different chronic-degenerative diseases [9]. The United States dominates the international fruit and vegetable trade market and is the first place in terms of imports and exports of food of plant origin, while the European Union, as a whole, is the second importer and exporter of this type of food. On the other hand, some Latin American countries, such as Chile, have become one of the main suppliers in the international fresh fruit market. The increase in the production and commercialization of fruits and vegetables has been increasing, reaching a global production of nearly 1 billion tons of fruits and vegetables [10]. Mexico ranks sixth in the fruit-producing countries, along with China, India, Brazil, the United States, and Italy. China, India, and Brazil concentrate about 30% of the total production of fruits worldwide; however, much of this production is destined for local consumption, so the impact on the world market is minimal [10, 11]. The low international trade attends to the lack of fruit conservation mechanisms, which can suffer damage during the transfer. This implies significant postharvest losses, thus reducing the possibilities of export and international marketing of a large number of vegetable crops.

3. Postharvest losses: causes and consequences

Currently, the losses of food in the world are about 1300 million tons per year; of which, in Latin America, the loss is, on average, 127 million tons of food per year; and with respect to fruits and vegetables, their loss is up to 55% in Latin America [12]. During the postharvest handling of fruits and vegetables, the losses range from 25 to 60% of the total production. This is due to several factors, but one of the most important factors is the poor handling of the products, where mechanical damage and diseases caused by pathogens play an important role [13]. Nowadays, the application of chemical fungicides is widely used as a strategy to control postharvest pathogens; however, even when they are effective, the environmental impact of these methods is negative since they are toxic. On the other hand, there are several reports about the adaptation of pathogens (resistance) that reduce the antifungal activity and difficult their control; thus, it is necessary to investigate new, secure, and effective alternatives for controlling postharvest diseases [13, 14]. Fungal contamination can occur during the handling of fruits and vegetables at field, postharvest handling, storage, and transport and cause the deterioration of products, and as a consequence, decrease the amount of available products and affect the financial benefits [15].

4. Alternative methods for controlling postharvest diseases

4.1 Physical treatments

Heat treatments are one of the physical methods that have been used most for the conservation of minimally processed postharvest fruits, either alone or combined with other eco-friendly alternatives [15]. Heat treatments have been applied in several fruits against different pathogens with good results, like peach (*Monilinia fructicola*), apple (*Penicillium expansum*), grapefruit (*P. digitatum*), sweet cherry (*P. expansum*), table grapes (*Botrytis cinerea*), among others [16].

4.2 Essential oils

The use of essential oils to control postharvest diseases is gaining popularity due to its safety features, biodegradability, and that are eco-friendly compounds [17]. For a long time, it has been recognized that some essential oils have antimicrobial, antiviral, antifungal, antiparasitic, and insecticidal properties [17]. Besides, in order to protect them and favoring their efficacy for controlling postharvest pathogens, the essential oils can be combined with edible films and coating, as previously reported with good results [17–19].

4.3 Edible films and coatings

The use of edible films and coatings is an alternative to the use of fungicides to preserve the postharvest quality of fruits and vegetables [20]. The application of coatings and edible films in foods is mainly in perishable products, such as horticultural products, due to their properties such as cost, availability, functional attributes, mechanical properties (flexibility, tension), optical properties (brightness and opacity), the barrier effect against the flow of gases, structural resistance to water and microorganisms, as well as sensory acceptability [21]. Besides, edible films and coatings have a high potential to incorporate active ingredients such as antibrowning agents, colorants, flavors, nutrients, spices, antimicrobial compounds, and antagonists; which may favor extending shelf life of the product and reducing the risk of pathogen growth [22]. In this sense, incorporation of different additives into polymeric matrices in several fruits coated has been evaluated successfully in controlling postharvest diseases and maintaining the fruit quality [18, 19, 23–25].

5. Biological control at postharvest stage

There is an international tendency to reduce the use of fungicides in fruit and vegetable products and to develop safer alternatives that reduce postharvest deterioration in fruits [26–28]. One option is the development of alternative control systems for controlling postharvest pathogens on fruits: as physical treatments (hot water), application of substance of vegetable and animal origin, or the use of antagonistic microorganisms [2, 29, 30]. Biological control is a promising alternative with high application potential for controlling postharvest pathogens. Some of the advantages of biological control compared to the use of fungicides are the absence of toxic waste in the fruits, friendly relationship with the environment, as well as in safety and health for the people who handle the products [14]. For several decades, a large number of microorganisms with potential characteristics of antagonistic organisms have been isolated from different sources (leaves, fruits, marine environment) [4, 31]. Bacteria, yeasts, and fungi are the main antagonistic organisms that have been used at postharvest stage. Countries such as the United States, South Africa, Spain, Italy, and Israel, among others, have developed and, in some cases, commercialized antagonistic microorganisms for their commercial use in postharvest, for the control of different diseases caused by the main pathogenic fungi of fruits of temperate, tropical, and subtropical origin [27, 32, 33]. The most widely used antagonists have an outstanding capacity for rapid growth and colonization of the fruit, nutrient acquisition, ability to withstand extreme temperature conditions, solar radiation, control capacity at low concentrations, as well as genetic stability and development capacity in economic cultivation means, ease of application in the management of postharvest. The main mechanisms of action present in the antagonists are antibiosis, production of lytic enzymes, competence by nutrients and space, or the induction of resistance mechanisms [1, 34].

There is a large number of reports of *in vitro* and *in vivo* tests, using bacteria such as *Pseudomonas cepacia* Van Hall, *Pseudomonas syringae*, *Bacillus subtilis*; as well as yeasts such as *Candida sake*, *Rhodotorula glutinis, Debaryomyces hansenii* for the control of the main pathogenic fruit fungi in postharvest, such as *Penicillium digitatum*, *Penicillium italicum, Alternaria, Aspergillus, Botrytis, Fusarium, Geotrichum, Gloeosporium, Monilinia, Penicillium, Mucor, Colletotrichum*, and *Rhizopus* [4, 27, 32, 33, 35]. In addition to bacteria and yeasts used in the biological control postharvest pathogens, the application of *Trichoderma*, alone or in combination with other alternative control systems in different fruits for the control of pathogenic fungi of fruits in the fruit, has been evaluated with favorable results [36, 37].

6. *Trichoderma*: application on fruits

Trichoderma has been applied as postharvest biocontrol agent in different crops such as papayas, strawberries, tomatoes, apples, pears, and bananas (**Table 1**). Existing different species of *Trichoderma* with high antagonistic capacity are *T. asperellum*, *T. viride*, and *T. harzianum*. Several authors have investigated different species of *Trichoderma* with the objective to find the most effective biocontrol agent for each crop and pathogen. The efficacy of six species of *Trichoderma* was evaluated on papaya fruit reducing the diameter lesions and disease incidence caused by *Colletotrichum gloeosporioides* by the application of *T. asperellum* and *T. viride* [6]. In a similar study on mango fruit, good efficacy was reported for controlling anthracnose applying *T. harzianum* obtaining a lower disease incidence (41.7% disease incidence) in fruits compared to control [38]. At present, *Trichoderma* is produced at industrial level as active component of biological products “biopesticides”; other ingredients that conform the biopesticides are the edible polymers, which can form coating for easy adhesion to the fruit and give to the product protection and stability during its shelf life. The application of biopesticides is widely used in agriculture and can be applied by immersion or spraying during the industrialization of

Postharvest fruit	Disease	Pathogen	*Trichoderma*species	Disease inhibition (%)	Reference
Banana	Crown rot	*Lasiodiplodia theobromae*	*T. viride*	60.7	Mortuza and Ilag [43]
			T. harzianum	56.2	
Pear	Rotten spots	*Rhizopus stolonifer*	*T. harzianum*	62.9	Batta [36]
	Gray mold	*Botrytis cinerea*		5.3	
	Blue mold	*Penicillium expansum*		33.3	
Mango	Antracnosis	*Colletotrichum gloeosporioides*	*T. harzianum*	60	Prabakar et al. [38]
Apple	Blue mold	*Penicillium expansum*	*T. atroviride* cepa P1	70.6	Quaglia et al. [44]
			T. harzianum Rifai cepa T22	72.5	
			T. harzianum Rifai cepa T67	70.6	
			T. reesei cepa T34	71.8	
			Trichoderma spp. cepa 8009	59.3	
Apple	Gray mold	*Botrytis cinerea*	*T. harzianum*	50.5	Batta [40]
	Blue mold	*Penicillium expansum*		40.6	
Pear	Gray mold	*Botrytis cinerea*		28.7	
	Blue mold	*Penicillium expansum*		47.6	
Papaya	Anthracnose	*Colletotrichum gloeosporioides*	*T. longibrachiatum*	65.0	Valenzuela et al. [6]
			T. viride	77.3	
Tomato	Gray mold	*Botrytis cinerea*	*T. harzianum*	96.9	Dal Bello et al. [45]

Table 1.
Percentage of inhibition of different postharvest diseases by the application of different species of Trichoderma.

agricultural products [39]. The incorporation of *T. harzianum* into edible coatings as biopesticide produced higher inhibition of the pathogens *Botrytis cinerea* and *Penicillium expansum* compared to the application as simple conidial suspension of the antagonist on the fruits [40]. The same author previously reported the same effect on other fruits such as pears, grapes, apples, strawberries, kiwis, and peaches [36, 41]. Microbial antagonists not only have antifungal properties, but also can act as inductor agents; some studies report that their application can induce biochemical defense responses by the interaction antagonist-fruit. In this sense, the application of *T. virens* decreases the blue mold incidence of apple fruits caused by *Penicillium expansum* by an increase in the enzymatic activity of peroxidase, catalase, β-1,3-glucanase, and the concentration of phenolic compounds, related as defense mechanisms against pathogens [42]. Studies realized in different plants such as tobacco, broccoli, tomato, lemon, apple, potato, and rice, reported that different *Trichoderma* species promote the expression of genes dependent on the defense mechanism of plants. Some of the expressed genes were chit36, chit42, agn13.1, and gluc78, which correspond to defense enzymes against cellular attack [7].

7. *Trichoderma*: mechanisms for controlling pathogens

The biocontrol mechanisms attributed to *Trichoderma* spp. are: competition for nutrients, parasitism, antibiosis, secretion of enzymes, and the production of inhibitor compounds [46, 47]. This biocontrol agent attacks and penetrates fungal cells, causing an alteration with the consequent degradation of the cell wall, causing retraction of the plasma membrane and disorganization of the cytoplasm [48]. These mechanisms are favored by the ability of *Trichoderma* to colonize the rhizosphere of plants.

7.1 Competition

Competition is defined as the unequal behavior of two or more organisms before the same requirement (substrate, nutrients), if the use of this substrate by one of the organisms reduces the amount or space available to others. This type of antagonism is favored by the characteristics of the biological control agent as ecological plasticity, growth rate primarily as chlamydospores [49], speed of development, and external factors such as soil type, pH, temperature, and humidity [50]. Nutrient competition can occur for nitrogen nonstructural carbohydrates (sugars and polysaccharides such as starch, cellulose, chitin, laminarin, and pectin, among others) and microelements.

7.2 Mycoparasitism

Mycoparasitism is defined as an antagonistic symbiosis between organisms, generally involving extracellular enzymes such as chitinases, cellulases, and which correspond to the composition and structure of the cell walls of parasitized fungi [51]. *Trichoderma* species' mycoparasitism during chemotropical process grows toward the host; hyphae adhere to it, wound on them frequently, and sometimes penetrate. Degradation of the cell walls of the host is observed in the late stages of the parasitic process [52], which leads to almost total phytopathogen weakening. This process is explained in four stages, within which it is recognized [53]:

1. *Chemotrophic growth*: the positive chemotropism directs growth toward a chemical stimulus.

2. *Recognition:* it is based on lectin-carbohydrate interactions. The lectins are proteins linked to sugars or glycoproteins, which agglutinate cells and are involved in the interactions between the components of the cell surface and their extracellular environment.

3. *Adhesion and curl:* occurs when the acknowledgment response is positive, *Trichoderma* hyphae adhere to host-mediated enzymatic processes. Hyphae adhesion occurs through association of a sugar antagonist wall with a lectin present in the wall of the pathogen.

4. *Lytic activity:* this stage is the production of extracellular lytic enzymes, mainly chitinases, glucanases, and proteases, which degrade the cell walls of the host and allow the penetration of antagonist hyphae.

Trichoderma can excrete metabolites like cellulases, glucanases, lipases, proteases, and chitinases in order to facilitate the insertion of hyphae for nutrient uptake of the pathogen, ending with the loss of cytoplasmic contents of the host cell [54]. The remaining cytoplasm is mainly surrounding the invading hyphae, showing signs of disintegration.

7.3 Antibiosis

Antibiosis is the inhibition of pathogen development by metabolized products and small toxic molecules, volatile and lytic enzymes, which operate structural polymers, such as chitin and β-1-3-glucans of the cell wall in most pathogenic fungi, producing an adverse effect on development and differentiation [55]. Given the above, it is said that the greater the amount of metabolic products, the antagonistic power increases; additionally, some authors mention that this mechanism is not the principle, due to the risk of emergence of the antibiotic-resistant pathogens [55].

7.4 Secretion of enzymes

The production of enzymes such as chitinase and/or glucanases produced by the fungus of *Trichoderma* is involved in the control of pathogenic fungi. These hydrolytic enzymes can degrade the cell wall polysaccharides (chitin and β glucans) affecting its stability and integrity [5].

7.5 Production of inhibitor compounds

The ability to induce resistance in a wide range of diseases caused by various classes of pathogens (including fungi, bacteria, and viruses) in a wide variety of plants may be an important characteristic of *Trichoderma* [56]. Currently, three classes of compounds are known that are produced by *Trichoderma* strains and that induce resistance in the plant. These are proteins with enzymatic functions, homologs of proteins encoded for avirulence (Avr), and oligosaccharides.

8. Advantages and limitations on the use of *Trichoderma*

Fungi possess characteristics that define their potential as biocontrol agents. *Trichoderma* species are cosmopolitan microorganims, inhabitant natural of soil such as organic matter, decaying wood as well as in crops waste. *Trichoderma* has several advantages as a biological control agent, such as a rapid growth and development as well as good production of a large number of enzymes inducible with the presence of phytopathogenic fungi [57]. In soil, Trichoderma has the ability to assimilate nutrients faster than pathogens, favoring its establishment and development, thus controlling pathogen infection and dissemination. Its use as a biocontrol agent can provide excellent advantages from the economic, environmental, and biological point of view, since they do not cause deterioration to the environment, do not affect the development of the plants, their production is cheaper, and its use does not entail the emergence of new pests or secondary pests [58]. However, its production on an industrial scale has some drawbacks that have limited the development of these organisms with wide possibilities as antagonist [59]. Even when, *Trichoderma* already exists in commercial form, its storage life is short, several investigations have been carried out finding that the best methods of fungus conservation [60]. In this sense, the incorporation in the formulations of different additives can improve microbial viability [2]. Other important limitations in the use of *Trichoderma* involve the lack of precise information to farmers, the little training about how to use the antagonistic fungi, as well as the lack of government economic support [61].

9. *Trichoderma* in combination with other control systems at postharvest management

In the management of postharvest fruit diseases, the use of antagonistic fungi is an alternative with great potential; specifically, the application of *Trichoderma* and its various species has obtained very promising results leading to the fabrication of commercial biocontrol products [62]. In addition, considering their mechanisms of action, such as competition for nutrients, production of lytic enzymes, parasitism, antibiosis, and induction of resistance, ensures the control and destruction of the main postharvest fungi. Even when *Trichoderma* has several mechanisms of action reported [63–65], the sole application of antagonists for controlling postharvest pathogens does not ensure 100% of disease control, that is why the development of new control technologies like combined system with other substances and methods could offer economical treatments and enhance biocontrol activity [66–68]. The combination of strains of *Pseudomonas-Trichoderma* and *Trichoderma viride-Bacillus subtilis* was effective controlling *in vitro* and *in vivo* tests such as *Penicillium digitatum* and *Penicillium expansum* and *Fusarium moniliforme,* respectively, in citrus and grapes. These studies confirm the synergistic effect of the combination of biological control agents [69, 70]. The addition of *Trichoderma harzianum* in polymeric matrices like chitosan has been evaluated against *Fusarium oxysporum* with good results, inducing defense mechanisms and the production of lytic enzymes as well as parasitism. In strawberry fruits, the incorporation of *Trichoderma* in chitosan and the application of a physical treatment (hot air) were effective to reduce microorganisms and maintain the fruit quality [71, 72]. The efficacy of the combination of bacterial (*P. syringae*) and fungal (*Trichoderma*) antagonists has been evaluated against *Botrytis cinerea* and *Fusarium oxysporum in vitro* tests, a synergistic effect was observed controlling successfully both fungus by the alteration and, consequently, the degradation of cell wall [73]. It has been reported that the different species of *Trichoderma* present different degrees of pathogen control at postharvest stage applied individually, but if the strains are combining, the potential of biocontrol increases, as previously reported in combinations with *T. viride*, *T. harzianum,* and *T. koningii* in the control of *Colletotrichum musae* reducing the incidence of crown rot [68]. The use of GRAS substances like sodium bicarbonate with *Trichoderma harzianum* was effective to control crown rot in banana fruits (*Colletotrichum musae* and *Fusarium verticillioides*) [74]. Recently, silver nanoparticles with *Trichoderma* have been synthesized and their antifungal capacity against postharvest pathogenic fungi such as *Alternaria*, *Penicillium*, and *Fusarium* has been evaluated; a good control on the mycelial growth and development of the pathogenic fungi was reported at low concentrations of *Trichoderma* spp. [75].

10. Conclusions

Trichoderma is a biocontrol agent with several benefits for its application on different commodities; this fungus represents a suitable eco-friendly alternative to fungicides by reducing postharvest losses.

Conflict of interest

Authors declare no conflict of interest.

Author details

Ramsés González-Estrada[1*], Francisco Blancas-Benítez[1], Beatriz Montaño-Leyva[2], Cristina Moreno-Hernández[1], Luz Del Carmen Romero-Islas[1], Jovita Romero-Islas[1], Rosa Avila-Peña[3], Anelsy Ramos-Guerrero[1], Angel Fonseca-Cantabrana[1] and Porfirio Gutierrez-Martinez[1]

1 National Technology of Mexico/Technological Institute of Tepic, Tepic, Nayarit, Mexico

2 University of Sonora, Hermosillo, Sonora, Mexico

3 Technological University of Nayarit, Tepic, Nayarit, Mexico

*Address all correspondence to: ramgonzalez@ittepic.edu.mx

References

[1] Nunes C. Biological control of postharvest diseases of fruit. European Journal of Plant Pathology. 2012;**133**:181-196. DOI: 10.1007/s10658-011-9919-7

[2] Sharma R, Singh D, Singh R. Biological control of postharvest diseases of fruits and vegetables by microbial antagonists: A review. Biological Control. 2009;**50**:205-221. DOI: 10.1016/j.biocontrol.2009.05.001

[3] Gutiérrez-Martínez P, Ramos-Guerrero A, Rodríguez-Pereida C, Coronado-Partida L, Angulo-Parra J, González-Estrada R. Chitosan for postharvest disinfection of fruits and vegetables. In: Wasim Siddiqui M, editor. Postharvest Disinfection of Fruits and Vegetables. 1st ed. Academic Press: Elsevier; 2018. pp. 231-241. DOI: 10.1016/B978-0-12-812698-1.00012-1

[4] Medina-Cordova N, Rosales-Mendoza S, Hernández-Montiel L, Angulo C. The potential use of *Debaryomyces hansenii* for the biological control of pathogenic fungi in food. Biological Control. 2018;**121**:216-222. DOI: 0.1016/j.biocontrol.2018.03.002

[5] Howell C. Mechanisms employed by *Trichoderma* species in the biological control of plant diseases: The history and evolution of current concepts. Plant Disease. 2003;**87**:4-10. DOI: 0.1094/PDIS.2003.87.1.4

[6] Valenzuela N, Angel D, Ortiz D, Rosas R, García C, Santos M. Biological control of anthracnose by postharvest application of *Trichoderma* spp. on maradol papaya fruit. Biological Control. 2015;**91**:88-93. DOI: 0.1016/j.biocontrol.2015.08.002

[7] Singh A, Shukla N, Kabadwal B, Tewari A, Kumar J. Review on plant-*Trichoderma*-pathogen interaction. International Journal of Current Microbiology and Applied Sciences. 2018;**7**:2382-2397. DOI: 10.20546/ijcmas.2018.702.291

[8] W. H. Organization. Fruit and Vegetable Promotion Initiative: A Meeting Report, 25-27/08/03. Geneva: World Health Organization; 2003

[9] W. H. Organization. Diet, Nutrition, and the Prevention of Chronic Diseases: Report of a Joint WHO/FAO Expert Consultation. Vol. 916. World Health Organization; 2003. Available from: http://apps.who.int/iris/bitstream/handle/10665/42665/?sequence=1. [Review date: November 2018]

[10] Barbosa-Cánovas G. Handling and Preservation of Fruits and Vegetables by Combined Methods for Rural Areas: Technical Manual, No. 149. Food and Agriculture Organization; 2003. Available from: https://books.google.com.mx/books?hl=es&lr=&id=IKsaRY9sb_AC&oi=fnd&pg=PA1&dq=Handling+and+P reservation+of+Fruits+and+Vegetabl es+by+Combined+Methods+for+Rural+Areas&ots=8RHV2OIVqt&sig=0dHL3i44Vkk2mK5gycYDAQNwA3E#v=onepage&q=Handling%20and%20Preservation%20of%20Fruits%20and%20Vegetables%20by%20Combined%20Methods%20for%20Rural%20Areas&f=false. [Review date:November 2018]

[11] FAOSTAT. "Agriculture Organization of the United Nations Statistics Division (2014)," Prod. 2016. Available from: http//faostat3.fao.org/browse/Q/QC/S [Review date: April 2015]

[12] FAO. Losses and food waste in Latin America and the Caribbean, 2018. Available from: http://www.fao.org/americas/noticias/ver/en/c/239392/. [Review date: November 2018]

[13] Janisiewicz W, Korsten L. Biological control of postharvest diseases of fruits. Annual Review of Phytopathology. 2002;**40**:411-441. DOI: 0.1146/annurev.phyto.40.120401.130158

[14] Droby S, Wisniewski M, El-Ghaouth A, Wilson C. Biological control of postharvest diseases of fruit and vegetables: Current achievements and future challenges. In: XXVI International Horticultural Congress: Issues and Advances in Postharvest Horticulture. 2002. pp. 703-713. DOI: 10.17660/ActaHortic.2003.628.89

[15] Sui Y, Wisniewski M, Droby S, Norelli J, Liu J. Recent advances and current status of the use of heat treatments in postharvest disease management systems: Is it time to turn up the heat? Trends in Food Science and Technology. 2016;**51**:34-40. DOI: 0.1016/j.tifs.2016.03.004

[16] Wisniewski M, Droby S, Norelli J, Liu J, Schena L. Alternative management technologies for postharvest disease control: The journey from simplicity to complexity. Postharvest Biology and Technology. 2016;**122**:3-10. DOI: 0.1016/j.postharvbio.2016.05.012

[17] Sivakumar D, Bautista-Baños S. A review on the use of essential oils for postharvest decay control and maintenance of fruit quality during storage. Crop Protection. 2014;**64**:27-37. DOI: 0.1016/j.cropro.2014.05.012

[18] González-Estrada R, Calderón-Santoyo M, Ragazzo-Sánchez J, Peyron S, Chalier P. Antimicrobial soy protein isolate-based films: Physical characterisation, active agent retention and antifungal properties against *Penicillium italicum*. International Journal of Food Science and Technology. 2018;**53**:921-929. DOI: 0.1111/ijfs.13664

[19] González-Estrada R, Chalier P, Ragazzo-Sánchez J, Konuk D, Calderón-Santoyo M. Antimicrobial soy protein based coatings: Application to Persian lime (*Citrus latifolia* Tanaka) for protection and preservation. Postharvest Biology and Technology. 2017;**132**:138-144. DOI: 0.1016/j.postharvbio.2017.06.005

[20] Han J. Edible films and coatings: A review. In: Han J, editor. Innovations in Food Packaging. 2nd ed. Academic press, Elsevier; 2014. pp. 213-255. DOI: 0.1016/B978-0-12-394601-0.00009-6

[21] Falguera V, Quintero J, Jiménez A, Muñoz J, Ibarz A. Edible films and coatings: Structures, active functions and trends in their use. Trends in Food Science and Technology. 2011;**22**:292-303. DOI: 0.1016/j.tifs.2011.02.004

[22] Otoni C, Avena-Bustillos R, Azeredo H, Lorevice M, Moura M, Mattoso L, et al. Recent advances on edible films based on fruits and vegetables—A review. Comprehensive Reviews in Food Science and Food Safety. 2017;**16**:1151-1169. DOI: 0.1111/1541-4337.12281

[23] González-Estrada R, Carvajal-Millán E, Ragazzo-Sánchez J, Bautista-Rosales P, Calderón-Santoyo M. Control of blue mold decay on Persian lime: Application of covalently cross-linked arabinoxylans bioactive coatings with antagonistic yeast entrapped. LWT- Food Science and Technology. 2017;**85**:187-196. DOI: 0.1016/j.lwt.2017.07.019

[24] Ponce A, Roura S, Del Valle C, Moreira M. Antimicrobial and antioxidant activities of edible coatings enriched with natural plant extracts: in vitro and in vivo studies. Postharvest Biology and Technology. 2008;**49**:294-300. DOI: 0.1016/j.postharvbio.2008.02.013

[25] Chávez-Magdaleno M, González-Estrada R, Plascencia-Jatomea M, Gutiérrez-Martínez P. Effect of pepper tree (Schinus molle) essential

oil-loaded chitosan bio-nanocomposites on postharvest control of *Colletotrichum gloeosporioides* and quality evaluations in avocado (*Persea americana*) cv. Hass. Food Science and Biotechnology. 2018;**27**: 1871-1875. DOI: 0.1007/s10068-018-0410-5

[26] Liu J, Sui Y, Wisniewski M, Droby S, Liu Y. Utilization of antagonistic yeasts to manage postharvest fungal diseases of fruit. International Journal of Food Microbiology. 2013;**167**:153-160. DOI: 0.1016/j.ijfoodmicro.2013.09.004

[27] Droby S, Wisniewski M, Macarisin D, Wilson C. Twenty years of postharvest biocontrol research: Is it time for a new paradigm? Postharvest Biology and Technology. 2009;**52**:137-145. DOI: 0.1016/j.postharvbio.2008.11.009

[28] Mari M, Bautista-Baños S, Sivakumar D. Decay control in the postharvest system: Role of microbial and plant volatile organic compounds. Postharvest Biology and Technology. 2016;**122**:70-81. DOI: 10.1016/j.postharvbio.2016.04.014

[29] Gutierrez-Martinez P, Ledezma-Morales A, Romero-Islas L, Ramos-Guerrero A, Romero-Islas J, Rodríguez-Pereida C, et al. Antifungal activity of chitosan against postharvest fungi of tropical and subtropical fruits. In: Dongre R, editor. Chitin-Chitosan-Myriad Functionalities in Science and Technology. 1st ed. Rijeka: IntechOpen; 2018. pp. 311-327. DOI: 10.5772/intechopen.76095

[30] Gutiérrez-Martínez P, Osuna-López SG, Calderón-Santoyo M, Cruz-Hernández A, Bautista-Baños S. Influence of ethanol and heat on disease control and quality in stored mango fruits. LWT- Food Science and Technology. 2012;**45**:20-27. DOI: 0.1016/j.lwt.2011.07.033

[31] Kong Q. Marine microorganisms as biocontrol agents against fungal phytopathogens and mycotoxins. Biocontrol Science and Technology. 2018;**28**:77-93. DOI: 0.1080/09583157.2017.1419164

[32] Liu J, Wisniewski M, Artlip T, Sui Y, Droby S, Norelli J. The potential role of PR-8 gene of apple fruit in the mode of action of the yeast antagonist, *Candida oleophila*, in postharvest biocontrol of *Botrytis cinerea*. Postharvest Biology and Technology. 2013;**85**:203-209. DOI: 0.1016/j.postharvbio.2013.06.007

[33] Calvo-Garrido C, Viñas I, Usall J, Rodríguez-Romera M, Ramos M, Teixidó N. Survival of the biological control agent *Candida* sake CPA-1 on grapes under the influence of abiotic factors. Journal of Applied Microbiology. 2014;**117**:800-811. DOI: 0.1111/jam.12570

[34] Spadaro D, Droby S. Development of biocontrol products for postharvest diseases of fruit: The importance of elucidating the mechanisms of action of yeast antagonists. Trends in Food Science and Technology. 2016;**47**:39-49. DOI: 0.1016/j.tifs.2015.11.003

[35] González-Estrada R, Ascencio-Valle F, Ragazzo-Sánchez J, Calderón-Santoyo M. Use of a marine yeast as a biocontrol agent of the novel pathogen *Penicillium citrinum* on Persian lime. Emirates Journal of Food and Agriculture. 2017;**29**:114-122. DOI: 0.9755/ejfa.2016-09-1273

[36] Batta Y. Control of postharvest diseases of fruit with an invert emulsion formulation of *Trichoderma harzianum* Rifai. Postharvest Biology and Technology. 2007;**43**:143-150. DOI: 0.1016/j.postharvbio.2006.07.010

[37] Wilson-Wijeratnam R, Marikar F, Abeyesekere M, Wijesundera R, Sivakumar D. Antagonistic effect of *Trichoderma harzianum* on post

harvest pathogens of rambutans. In: IV International Conference on Postharvest Science. 2000. pp. 389-392. DOI: 10.17660/ActaHortic.2001.553.88

[38] Prabakar K, Raguchander T, Saravanakumar D, Muthulakshmi P, Parthiban V, Prakasam V. Management of postharvest disease of mango anthracnose incited by *Colletotrichum gloeosporioides*. Archives of Phytopathology and Plant Protection. 2008;**41**:333-339. DOI: 0.1080/03235400600793502

[39] Marín A, Atarés L, Chiralt A. Improving function of biocontrol agents incorporated in antifungal fruit coatings: A review. Biocontrol Science and Technology. 2017;**27**:1220-1241. DOI: e0.1080/09583157.2017.1390068

[40] Batta Y. Production and testing of biopesticide for control of postharvest mold infections on fresh fruits of apple and pear. Advances in Microbiology. 2015;**5**:787. DOI: 0.4236/aim.2015.512083

[41] Batta YA. Effect of treatment with *Trichoderma harzianum* Rifai formulated in invert emulsion on postharvest decay of apple blue mold. International Journal of Food Microbiology. 2004;**96**:281-288. DOI: 0.1016/j.ijfoodmicro.2004.04.002

[42] Bordbar F, Etebarian H, Sahebani N, Rohani H. Control of postharvest decay of apple fruit with *Trichoderma virens* isolates and induction of defense responses. Journal of Plant Protection Research. 2010;**50**:146-152. DOI: 0.2478/v10045-010-0025-1

[43] Mortuza M, Ilag L. Potential for biocontrol of *Lasiodiplodia theobromae* (Pat.) Griff. & Maubl. in banana fruits by *Trichoderma* species. Biological Control. 1999;**15**:235-240. DOI: 0.1006/bcon.1999.0716

[44] Quaglia M, Ederli L, Pasqualini S, Zazzerini A. Biological control agents and chemical inducers of resistance for postharvest control of *Penicillium expansum* Link. on apple fruit. Postharvest Biology and Technology. 2011;**59**:307-315. DOI: 0.1016/j.postharvbio.2010.09.007

[45] Dal Bello G, Lampugnani G, Abramoff C, Fusé C, Perelló A. Postharvest control of *Botrytis* gray mould in tomato by antagonists and biorational compounds. IOBC-WPRS Bulletin. 2015;**111**:417-425

[46] Guédez C, Cañizález L, Castillo C, Olivar R. Efecto antagónico de *Trichoderma harzianum* sobre algunos hongos patógenos postcosecha de la fresa (Fragaria spp). Revista de la Sociedad Venezolana de Microbiología. 2009;**29**:34-38

[47] Zimand G, Elad Y, Chet I. Effect of *Trichoderma harzianum* on *Botrytis cinerea* pathogenicity. Phytopathology. 1996;**86**:1255-1260

[48] Tronsmo A, Raa J. Antagonistic action of *Trichoderma pseudokoningii* against the apple pathogen *Botrytis cinerea*. Journal of Phytopathology. 1977;**89**:216-220

[49] Hjeljord L, Tronsmo A. *Trichoderma* and *Gliocladium* in bilogical control: A review. In: Harman GE, Kubicek CP, editors. *Trichoderma* & *Gliocladium*—Enzymes, Biological Control and Commercial Applications. 1st. ed. CRC Press, Taylor & Francis; 1998. pp. 131- 145

[50] Ahmad J, Baker R. Rhizosphere competence of *Trichoderma harzianum*. Phytopathology. 1987;**77**:182-189

[51] Lorito M, Harman G, Di Prieto A, Hayes C. Extracellular chitimolytie enzymes produced by *T. harzianum*, purification, characterization and molecular cloning. Phytopathology. 1990;**82**:10-77

[52] Carsolio C, Benhamou N, Haran S, Cortés C, Gutiérrez A, Chet I, et al. Role of the *Trichoderma harzianum* endochitinase gene, ech42, in mycoparasitism. Applied and Environmental Microbiology. 1999;**65**:929-935

[53] Mantyla A, Paloheimo M. Suominen Industrial mutants and recombinant strains of *Trichoderma* reesei. In: Harman GE, Kubicek CP, editors. *Trichoderma* & *Gliocladium*—Enzymes, Biological Control and Commercial Applications. 1st. ed. CRC Press, Taylor & Francis; 1998. pp. 291-304

[54] Demain A, Adrio J. Contributions of microorganisms to industrial biology. Molecular Biotechnology. 2008;**38**:41. DOI: 0.1007/s12033-007-0035-z

[55] Goldman G, Hayes C, Harman G. Molecular and cellular biology of biocontrol by *Trichoderma* spp. Trends in Biotechnology. 1994;**12**:478-482

[56] Harman G, Howell C, Viterbo A, Chet I, Lorito M. *Trichoderma* species—Opportunistic, avirulent plant symbionts. Nature Reviews. Microbiology. 2004;**2**:43

[57] Chávez García M. Producción de *Trichoderma* sp. y evaluación de suefecto en cultivo de crisantemo (Dendranthema grandiflora). 2006. Available from: https://repository.javeriana.edu.co/bitstream/handle/10554/8312/tesis286.pdf?sequence=1. [Review date:November 2018]

[58] Karas P, Perruchon C, Exarhou K, Ehaliotis C, Karpouzas D. Potential for bioremediation of agro-industrial effluents with high loads of pesticides by selected fungi. Biodegradation. 2011;**22**:215-228. DOI: 0.1007/s10532-010-9389-1

[59] Fernández-Larrea O. Los microorganismos en el control biológico. Alternativas de producción en Cuba. Memorias del Curso Int. Manejo Agroecol. Plagas en el Sist. Prod. Inisav. 2006

[60] Cárdenas YG. Métodos de conservación y formulación de *Trichoderma harzianum* Rifai. Fitosanidad. 2010;**14**:189-195

[61] Usall J, Torres R, Teixido N. Biological control of postharvest diseases on fruit: A suitable alternative? Current Opinion in Food Science. 2016;**11**:51-55. DOI: 0.1016/j.cofs.2016.09.002

[62] Woo S, Scala F, Ruocco M, Lorito M. The molecular biology of the interactions between *Trichoderma* spp., phytopathogenic fungi, and plants. Phytopathology. 2006;**96**:181-185. DOI: 0.1094/PHYTO-96-0181

[63] Brunner K, Zeilinger S, Ciliento R, Woo S, Lorito M, Kubicek C, et al. Improvement of the fungal biocontrol agent *Trichoderma atroviride* to enhance both antagonism and induction of plant systemic disease resistance. Applied and Environmental Microbiology. 2005;**71**:3959-3965. DOI: 10.1128/AEM.71.7.3959-3965.2005

[64] Whipps J, Lumsden R. Commercial use of fungi as plant disease biological control agents: Status and prospects. In: Butt TM, Jackson C, Magan N, editors. Fungi as Biocontrol Agents: Progress, Problems and Potential. Wallingford: CABI Publishing; 2001. pp. 9-22

[65] Marra R, Ambrosino P, Carbone V, Vinale F, Woo S, Ruocco M, et al. Study of the three-way interaction between *Trichoderma atroviride*, plant and fungal pathogens by using a proteomic approach. Current Genetics. 2006;**50**:307-321. DOI: 0.1007/s00294-006-0091-0

[66] Ramos-Guerrero A, González-Estrada R, Hanako-Rosas G, Bautista-Baños S, Acevedo-Hernández G, Tiznado-Hernández M, et al. Use of inductors in the control of *Colletotrichum gloeosporioides* and *Rhizopus stolonifer* isolated from soursop fruits: In vitro tests. Food Science and Biotechnology. 2018;**27**:755-763. DOI: 0.1007/s10068-018-0305-5

[67] Sivakumar D, Wijeratnam R, Abeyesekere M, Wijesundera R. Combined effect of generally regarded as safe (GRAS) compounds and *Trichoderma harzianum* on the control of postharvest diseases of rambutan. Phytoparasitica. 2002;**30**:43-51. DOI: 0.1007/BF02983969

[68] Sangeetha G, Usharani S, Muthukumar A. Biocontrol with *Trichoderma* species for the management of postharvest crown rot of banana. Phytopathologia Mediterranea. 2009;**48**:214-225. DOI: 10.14601/Phytopathol_Mediterr-2741

[69] Panebianco S, Vitale A, Polizzi G, Scala F, Cirvilleri G. Enhanced control of postharvest citrus fruit decay by means of the combined use of compatible biocontrol agents. Biological Control. 2015;**84**:19-27. DOI: 0.1016/j.biocontrol.2015.02.001

[70] Senthil R, Prabakar K, Rajendran L, Karthikeyan G. Efficacy of different biological control agents against major postharvest pathogens of grapes under room temperature storage conditions. Phytopathologia Mediterranea. 2011;**50**:55-64. DOI: 10.14601/Phytopathol_Mediterr-3115

[71] Nitu N, Masum M, Jannat R, Bhuiyan M, Sultana S. Application of chitosan and *Trichoderma* against soil-borne pathogens and their effect on yield of tomato (*Solanum lycopersicum* L.). International Journal of Biosciences. 2016;**9**:10-24

[72] Aml A, Ezzat A, Rageh M, Saber W, El-Sheikh T. Effect of chitosan, biocontrol agents and hot air to reduce postharvest decay and microbial loads of strawberries. Development. 2017;**6**:8

[73] Lorito M, Woo S, D'ambrosio M, Harman G, Hayes C, Kubicek C, et al. Molecular Plant-Microbe Interactions. 1996;**9**:206-213

[74] Alvindia D. Sodium bicarbonate enhances efficacy of *Trichoderma harzianum* DGA01 in controlling crown rot of banana. Journal of General Plant Pathology. 2013;**79**:136-144. DOI: 0.1007/s10327-013-0432-z

[75] Elamawi R, Al-Harbi R, Hendi A. Biosynthesis and characterization of silver nanoparticles using *Trichoderma longibrachiatum* and their effect on phytopathogenic fungi. Egyptian Journal of Biological Pest Control. 2018;**28**:28. DOI: 10.1186/s41938-018-0028-1

Chapter 5

A Review Report on the Mechanism of *Trichoderma* spp. as Biological Control Agent of the Basal Stem Rot (BSR) Disease of *Elaeis guineensis*

Syed Ali Nusaibah and Habu Musa

Abstract

Trichoderma spp. have been the most common fungi applied as biological control agents (BCA) as an effort to combat a wide range of plant diseases. Its uses have recorded good success rate in controlling major plant diseases. Knowledge on the mechanisms employed by *Trichoderma* spp. could be further studied to improve its ability as an efficient biocontrol agent. The *Trichoderma* ability to curb plant diseases were mainly based on the activation of single or multiple control mechanisms. It is known that the *Trichoderma*-based biocontrol mechanisms mainly rely on mycoparasitism, production of antibiotic and/or hydrolytic enzymes, competition for nutrients, as well as induced plant resistance; numerous secondary metabolites produced by *Trichoderma* species could directly inhibit the growth of several plant pathogens. These mechanisms may act directly or indirectly against the targeted plant pathogen. This chapter reviews the recent updates on published research findings on mechanisms used by *Trichoderma* as biological control of plant diseases particularly on basal stem rot disease of oil palm caused by *Ganoderma* spp.

Keywords: antibiosis, competition, induced resistance, mycoparasitism, secondary metabolite

1. Introduction

1.1 Biological control of oil palm basal stem rot disease caused by *Ganoderma boninense*

Many promising biological antagonists, mainly from *Trichoderma*, *Aspergillus*, *Penicillium*, *Pseudomonas*, and *Bacillus*, have been reported as effective antagonists against *Ganoderma boninense* in coconut [1] and oil palm [2–4]. In 1990, [5] evaluated the incorporation of *Trichoderma* spp. grown on dried palm oil mill effluent into planting holes as a prophylactic measure. Later, [6] reported delays in infection in the field following treatment with *Trichoderma*, but eventually, the disease incidence was similar to untreated controls. Thus, the possible explanation for this could be due to a low natural occurrence of *Trichoderma* spp. in the soil [7]. In Malaysia, certain *Trichoderma* strains such as *Trichoderma virens* and *Trichoderma harzianum* have shown good biocontrol ability against *G. boninense* in nurseries [8]. Besides that, in Indonesia, a biofungicide consisting of *Trichoderma koningii* was reported to reduce BSR in decomposing oil palm residues in the field [9].

Biological control of BSR disease in oil palm can be effectively achieved through the utilization of an effective strain of *Trichoderma* spp. The strain must not only have the potential mechanisms for biological control such as antibiosis and mycoparasitism but also a strong competitive ability to displace the causative fungus *G. boninense* so as to reduce the pathogen's opportunity for root colonization. It must be able to favorably compete and adopt well within the environment in which it will operate and be able to rapidly colonize and proliferate on the existing and newly formed roots immediately after its application [10].

2. Mechanisms of *Trichoderma* species

Trichoderma spp. employ several antagonistic mechanisms against plant pathogens. These include antibiosis, mycoparasitism, competition for nutrients and space, promotion of plant growth, induced plant defense mechanisms, and modification of environmental conditions [11].

2.1 Mycoparasitism

The potential of *Trichoderma* spp. to parasitize, suppress, or even kill other plant pathogenic fungi has been recognized as an important mechanism for its success as a biological control [12]. Mycoparasitism is a direct mechanism for biological control that works by parasitizing, detecting, growing, and colonizing pathogen [11, 13]. The ability to mycoparasitize other fungi has been widely used for the biological control of agricultural pests (mainly against pathogenic fungi and parasitic nematodes). Some species of *Trichoderma* such as *T. asperellum*, *T. atroviride*, *T. virens*, and *T. harzianum* are widely used as biological control agents of plant pathogens [11]. It is able to directly kill pathogens and other plant-associated fungi, with a wider host range in diverse ecologies [14]. This is done via the use of many mycoparasitic strategies [15, 16]. These mycoparasitic abilities appear to be very complex, involving the detection of plant pathogen through chemotropism; lysis of the pathogen's cell wall (the key to mycoparasitism) [17]; pathogen's hyphal penetration by appresorial formation; production of cell wall-degrading enzymes (CWDEs) and peptaibols, mediated by heterotrimeric G-proteins and mitogen-activated protein (MAP) kinases [11]; and parasitizing pathogen's cell wall contents [12]. Degradation of pathogen's cell wall during mycoparasitism is mediated by a set of hydrolytic enzymes including β-(1,6)-glucanases, chitinases, and proteases. Several members from each of these classes have been shown to be involved in mycoparasitism and/or to be induced under mycoparasitism-related growth conditions [18]. Genome analysis enabled the assessment of cell wall-degrading enzymes encoded in the genomes of *Trichoderma* spp. and unraveled even more complex enzymatic degradation machinery for fungal cell walls than previously anticipated [19].

Considerable research work has been done to identify and understand the enzymes induced by *Trichoderma* to recognize host pathogen [20]. The degradation of a pathogen's cell wall is an important aspect of mycoparasitism and biological control of plant diseases. *Trichoderma* also produces secondary metabolites (volatile and nonvolatile) [21]. With regard to the production of secondary metabolites, two

Trichoderma species—*T. virens* and *T. reesei*—are the highest producers of secondary metabolites [22].

Three important *Trichoderma* species—*T. virens*, *T. atroviride*, and *T. harzianum*—have been identified with the highest production of chitinolytic enzymes compared to other fungi known to have similar biological control abilities [21]. Most of the secondary metabolites' coding gene clusters are exclusively specific to certain *Trichoderma* spp., while some are found in all three species [21]. Studies on the responsible signal transduction pathways of *T. atroviride* during mycoparasitism have led to the isolation of key constituents of MAP and cAMP kinase signaling pathways, such as the α-subunits of G-proteins (G-α) that control antibiotic production, extracellular enzyme, and coiling around host hypha [23].

2.2 Promoting plant growth

Microbial organisms colonize a plant's root system and at the same time play a beneficial role in biological control, protecting the plant from soil-borne pathogens as well stimulating plant growth [13]. These beneficial relationships between plants and microbes often occur in the rhizosphere, improving plant growth or helping the plant overcome biotic or abiotic stresses [24]. *Trichoderma* spp. proliferate in the rhizosphere, establishing a symbiotic association, thus improving plant nutrition and growth in a natural way [25]. It is able to colonize roots, improving plant nutrition, growth, and development as well as enhancing plant resistance to abiotic stresses. Increasing plant growth by using biological control agents is usually attributed to an indirect effect associated with control of plant pathogens. It was also reported that the application of *T. harzianum* to cucumber or tomato seedlings increased the concentration of trace and essential elements such as Fe, Zn, Cu, Mn, Mg, Ca, N, P, K, and Na both in the shoots and roots [26]. This is due to its ability to produce many phytohormones, siderophores, and phosphate-solubilizing enzymes [27]. Phytohormones stimulate root growth, thus increasing the absorptive surface of plant roots. These phytohormones include cytokinins, indole-3-acetic acid, and gibberellins [28]. Growth promotion of plant antimicrobial compounds of *Trichoderma* has been demonstrated [29]. Harzianopyridone, 6PP, trichocereus A-D, koninginins, cyclonerodiol, harzianolide, and harzianic acid (HA) are examples of isolated compounds that promote plant growth in a concentration-dependent manner [30]. A novel secondary metabolite—cerinolactone—has been isolated and characterized from *Trichoderma cerinums* and was able to positively alter the growth of tomato seedlings 3 days after treatment [30]. Similarly, HA and iso-harzianic acid found in *T. harzianum* metabolites were also found to promote plant growth, through the strong binding of iron [30]. It was also shown that *T. virens* and *T. atroviride* produce certain types of indole-3-acetic acid-related indoles (IAA-related indoles). The production of IAA-related indoles in *Trichoderma* liquid cultures was stimulated by the addition of L-tryptophan. This observation proposed that the production of IAA-related indoles could be one of the mechanisms employed by *Trichoderma* to promote plant growth and an increase in the number of secondary roots, leading to a higher biomass in *Arabidopsis* [31]. On the other hand, [32] proposed that degradation of IAA by *Trichoderma* leads to a reduction of detrimental effects of IAA on root elongation that could cause reduced ethylene (ET) production, by decreasing its precursor, 1-aminocyclopropane-1-carboxylic acid (ACC), and/or ACC deaminase activity present in *Trichoderma*. Recently, it was shown that *T. asperellum* has high 1-aminocyclopropane-1-carboxylic acid deaminase (ACCD) activity when grown with ACC as the sole nitrogen source. The ACCD-encoding gene Tas-acdS was upregulated when ACC was added to the artificial growth medium. Silencing of Tas-acdS showed decreased ability of silenced transformants to promote root elongation of canola plants [33]. These mechanisms could be responsible for the plant growth-promoting activity of *Trichoderma* [31, 33]. The application of *Trichoderma* spp. results in significant vegetative growth on a wide range of crop plants [31, 34]. Interaction between *Trichoderma* spp. and the plant triggers enhanced immunity against plant diseases, thus improving plant health [35]. The ability to promote plant growth may be an important characteristic; however, this is not found in every *Trichoderma* species [13, 36]. Plant growth enhancement is evidenced by the increase in productivity, nutrient uptake, biomass, resistance to stress, and improvement of plant health [37]. *Trichoderma* isolates from the rhizosphere of the mangrove *Avicennia marina* solubilize P from the insoluble $Ca_3(PO_4)^2$ and correlate with an increase of the extracellular phytase activity—an acidic phosphatase and extracellular activity of phytase were induced only in the presence of $Ca_3(PO_4)^2$ [38]. In addition, the application of *Trichoderma* spp. in consortium was reported to enhance the physical strength and durability of the plant's cell wall toward cell wall-degrading plant pathogenic fungi [39–42]. It is likely that improvement in root growth was the effect of one or more mechanisms, and this may include an increase in soil nutrient solubilization, increase in the rate of carbohydrate photosynthetic activities and carbohydrate metabolism, plant growth regulatory effect, and increased rooting depth, thus increasing resistance to drought conditions [43]. *Trichoderma* spp. are more effective in colonizing and enhancing plant growth, if there are enriched inorganic soil substrates such as bioorganic fertilizers [44].

2.3 Induced resistance in plants

Apart from *Trichoderma*'s capacity to attack plant pathogens and inhibit its growth, many reports have shown the induction of systemic and local resistances against a wide range of pathogens [25, 45]. Both the inductions of systemic and local resistances in plants are the result of complex interactions between different elicitors released by microbes and plant receptors, which induces plant cells' physiological and biochemical changes. However, these processes are possible when the systemic acquired resistance (SAR) triggers a resistance in the entire foliage and enhances production of the defense signal molecule, salicylic acid (SA), for further signaling [46]. In addition, *Trichoderma* spp. were reported to induce the synthesis of regulatory proteins in plants especially under certain disease stress, where these regulatory proteins detect microbe effectors and activate the plant's defense systems [47]. Induction of systemic resistance by *Trichoderma* spp. has been scarcely studied, as compared to the response induced in plants by rhizobacteria, presumably because the research on *Trichoderma* focused on understanding the factors involved in antibiosis and mycoparasitism. Its ability to induce systemic and local resistances has been reported on both monocots and dicots [12, 31, 43, 45]. It is important to know that induced resistance varies from plant to plant due to the influence of environmental factors, the pathogens involved, and the symbionts' relationship in the rhizosphere [34, 48]. Plant's resistances are induced via the expression of pathogenesis-related (PR) genes mediated by SA and are known as the SAR, which is also triggered by biotrophic and/or hemibiotrophic pathogens [25].

The secondary metabolites 6-pentyl-α-pyrone and harzionalide produced by *T. atroviride* and *T. harzianum* induced systemic defense response in oilseed rape and tomato seedlings against *Leptosphaeria maculans* and *B. cinerea*. In addition, upregulation of PR genes was also found in plants treated with these secondary metabolites [49]. Expression analysis of SA-related genes in *Trichoderma* spp.-treated plants showed a long-lasting upregulation, possibly activating a priming mechanism in the plant. Inoculation of *Trichoderma*-treated plants with *B. cinerea* led to an enhanced expression of JA-related genes, triggering systemic resistance to the pathogen in a plant genotype-dependent mode [36]. Peptaibols also induce systemic resistance and defense responses in plants—*T. virens* produces peptaibols of 14 and 18 amino acids, with several isoforms each that induce systemic resistance [50]. In plant root colonization, *Trichoderma* spp. deal with a plant's defense system by synthesizing antimicrobial compounds such as phytoalexins. Its interactions with plants during the early stages of root colonization might stimulate the activation of cell detoxification and plant protection mechanisms [51].

2.4 Antibiosis

Trichoderma spp. are rich and important sources of secondary metabolites (SMs) used for biological control of plant diseases [12]. It was reported that antibiosis occurs during the interactions between a host plant, pathogens, and *Trichoderma* spp. that resulted in the production of antibiotics and low-molecular-weight compounds by *Trichoderma* to inhibit the growth of phytopathogenic fungi [52]. Fungal secondary metabolites also play important roles in interactions with plants [53]. Antimicrobial activities could be the result of several secondary metabolites such as peptaibols, terpenes, polyketides, gliotoxin, and gliovirin produced by fungi [49]. Other metabolites include tricholin, harzianic acid, viridian, gliosoprnins, heptelidic acid, 6-pentyl-α-pyrone, and massoilactone [54]. Secondary metabolites are classified into two major classes—useful and toxic compounds polyketide synthases and non-ribosomal peptide synthases [55] or volatile and nonvolatile secondary metabolites which deter colonization by competing with pathogenic fungi in ecological niches. Secondary metabolites are known for its ability to synthesize peptaibols. Peptaibols are from a family of peptides with antibiotic functions, characterized by C-terminal alcohol residues, short-chain-length amino acids (<20 residues), and high levels of nonstandard amino acids [21]. Peptaibols consist of 2-amino-isobutyric acid and other non-proteinogenic amino acids, produced as secondary metabolites with antibiotic activities against pathogenic fungi and bacteria. It has been discovered that peptaibols produced by *T. pseudokoningii* can induce programmed cell death in plant fungal pathogens, causing apoptotic deaths of the pathogens [56]. These are natural products biosynthesized by many fungi and work together with cell wall-degrading enzymes to inhibit or completely prevent the growth of pathogenic fungi and/or elicit development of induced plant resistance against pathogens [21]. Research has shown that these compounds demonstrated strong positive effects on plant growth and resistance in plants to abiotic and biotic stresses [57]. The production of antimicrobial compounds strongly depends on the availability of exogenous nutrients such as root exudates, leakage of nutrients on leaf surface, and/or organic nutrients in the soil, as well as environmental conditions [58]. Due to the advancement in analytical studies, many secondary metabolites from *Trichoderma* spp. have been isolated and identified. Over 120 *Trichoderma* secondary metabolites have been identified and reported by [59]. There are over 1000 *Trichoderma* peptaibol sequences known to date—some involved in biocontrol. This great potential of *Trichoderma* spp. to produce several types of secondary metabolites is reflected in the genomes of the three important species (http://genome.jgi-psf.org/) which amounted to 262 and 349 for *T. reesei* and *T. atroviride*, respectively. Most of the secondary metabolite genes present in *T. reesei* are also found in *T. atroviride* and *T. virens* [19]. Peptaibols—peptides containing a-aminoisobutyric acid (Aib) and a C-terminal 1,2-amino alcohol—are produced largely by members of *Trichoderma* [60]. Peptaibols have unusual amino acid content that resulted from non-ribosomal biosynthesis. Several functional enzymes known as peptide synthetases assemble these molecules via the multiple carrier thiotemplate mechanism from a remarkable range of precursors, which can be N-methylated, acylated, or reduced. Peptaibols show interesting biological and physiochemical properties that include the formation of pores in bilayer lipid membranes and antifungal, antibacterial, and rarely antiviral activities and elicit plant resistance [61]. These secondary metabolites induce certain reaction in plants when applied to the leaves or injected into roots or plant tissues. It also stimulates the biocontrol potential of *Trichoderma* by activating mycoparasitic gene expression which elicits defense mechanisms against plant pathogens [62].

Some exert antimicrobial effects at high doses, but others act as microbe-associated molecular patterns (MAMPs), while auxin-like compounds act at low concentrations. For example, even 1 ppm of 6-pentyl-α-pyrone, harzianolide, or harzianopyridone can activate a plant's defense system and regulate growth in plants (e.g., tomato, canola, and pea) [49]. Many elicitors released by *Trichoderma* that activate plant defense systems can be classified into several groups such as proteins or peptides [31], enzymes (cellulase, chitinase, glucanase, protease, or xylanases) [63, 64], and fatty acids, lipids, and its derivatives such as glycosphingolipids [12].

2.5 Competition

This is a classical mechanism of biological control of plant pathogens [65]. Competition among microorganisms occurs only when resources such as soil nutrients and space are limited. In this situation, antagonistic microbes produce secondary metabolites capable of inhibiting or slowing growth and other activities of pathogenic fungi, thus conferring ecological advantages over its competitors. Antagonistic microbes utilize available resources for growth, leaving pathogens with insufficient nutrients for its growth and consequently starved [66]. These antagonists favorably compete for iron causing the suppression of *Fusarium* spp. [67]. Iron is an essential mineral nutrient for many microbes. Iron in an aerobic condition (i.e., with oxygen and neutral pH) exists as Fe^{3+} in immobilized forms rather than as oxyhydroxides or hydroxides, thereby making it unavailable for microbes [68]. *Trichoderma* has a strong potential to mobilize and utilize soil nutrients, making it more efficient and competitive than many other soil microbes (fungi and bacteria). This process could be related to the production of inorganic acids, namely, citric, gluconic, and fumaric acids which decrease soil pH and increase the solubilization of phosphate and micronutrients (iron and manganese) [49]. *Trichoderma* secretes siderophores and is able to grow in conditions that are poor in iron by using residual immobilized Fe. Most fungi including *Trichoderma* produce various forms of siderophores, which help the fungi to overcome adverse soil conditions [68]. Siderophore production can be beneficial to the plant and microbes for two reasons: (1) siderophore production by antagonist fungi cansuppress the growth of plant pathogens by depriving it of an iron source and (2) siderophores can help in solubilizing iron that was unavailable to the plant [12]. Most of the siderophores isolated so far belong to the hydroxamate class and can be divided into three structural families: caprogens, ferrichromes, and fusarinines. Fungi typically produce more siderophores than other microbes [12]. Specifically, *T. harzianum* produces the highest amount of siderophores but does not have any unique compounds, while *T. reesei* can biosynthesize cis-fusarinine, a major siderophore. The variety of siderophore production in *Trichoderma* spp. is due to the modification of the non-ribosomal peptide synthetase (NRPS) products rather than diverse NRPS-encoding genes [69]. The ability of tetramic acid from *Trichoderma* spp. to bind with good affinity to Fe^{3+} explains the mechanism of iron solubilization that significantly changes nutrient levels in the soil for other microbes and the host plant. Siderophores produced by *Trichoderma* spp. have other effects and functions including virulence enhancement of pathogens, storage of intracellular iron, and suppression of microbe growth during competition with *Trichoderma* spp. [68].

Reduced nutrient concentrations lead to reduced conidial germination, slowed germ tube growth of a pathogen, reduced number of infection sites, and the extent of plant disease development [70]. *Trichoderma* has the ability to alleviate a wide range of abiotic and physiological stresses and also enhances plant nutrient uptake and increases nitrogen use efficiency. *Trichoderma* root colonization delayed the effects of drought-induced changes in stomatal conductance, green fluorescence emissions, and net photosynthesis resulting in an improved plant water status [71]. This may potentially be more important for plant disease control because *Trichoderma* deprives pathogens of available nutrients for growth and development, thus rendering it ineffective to cause any disease. Some *Trichoderma* species even demonstrated a potential to improve photosynthetic and respiratory activities of plants, resulting in its ability to reproduce plant gene expression, probably through the activation of a limited number of general plant pathways [25].

Author details

Syed Ali Nusaibah[1]* and Habu Musa[2]

1 Department of Plant Protection, Faculty of Agriculture, Universiti Putra Malaysia, Serdang, Selangor, Malaysia

2 Department of Crop Science, Faculty of Agriculture, Federal University Dutse, Dutse, Nigeria

*Address all correspondence to: nusaibah@upm.edu.my

References

[1] Wijesekera HTR, Wijesundera RLC, Rajapakse CNK. Hyphal interactions between *Trichoderma viride* and *Ganoderma boninense* Pat. cause of coconut root and bole rot. Journal of the National Science Council of Sri Lanka. 1996;**24**:217-219

[2] Dharmaputra OS, Tjitrosomo HS, Abadi AL. Antagonistic effect of four fungal isolates to *Ganoderma boninense*, the causal agent of basal stem rot of oil palm. Journal of Biotropia. 1989;**3**:41-49

[3] Ariffin D, Idris AS, Singh G. Status of Ganoderma in oil palm. In: Flood J et al., editors. Ganoderma Diseases of Perennial Crops. Wallingford, UK: CAB International Publishing; 2000. pp. 249-666

[4] Susanto A, Sudharto PS, Purba RY. Enhancing biological control of basal stem rot disease (*Ganoderma boninense*) in oil palm plantations. Mycopathologia. 2005;**159**:153-157

[5] Singh G. *Ganoderma*—The scourge of oil palm in the coastal areas. In: Ariffin D, Jalani S, editors. In: Proceedings of the *Ganoderma* Workshop, 11 September 1990. Bangi, Selangor, Malaysia: Palm Oil Research Institute of Malaysia; 1990. pp. 7-35

[6] Hasan Y, Turner PD. The comparative importance of different oil palm tissues as infection sources for basal stem rot in replantings. The Planter. 1998;**74**:119-135

[7] Sariah M, Zakaria H. The use of soil amends for the control of basal stem rot of oil palm seedlings. In: Flood et al., editors. Ganoderma Disease of Perennial Crops. UK: CABI Publishing; 2000. pp. 89-100

[8] Shamala S, Idris AS. Trichoderma as a biocontrol agent against Ganoderma in oil palm. MPOB Information Series No. 463. 2009. pp. 4

[9] Soepena H, Purba RY, Pawirosukarto S. A control strategy for basal stem rot (Ganoderma) on oil palm. In: Flood J, Bridge PD, Holderness M, editors. UK: CAB International; 2000. pp. 83-88

[10] Sariah M, Choo CW, Zakaria H, Norihan MS. Quantification and characterisation of *Trichoderma* spp. from different ecosystems. Mycopathologia. 2005;**159**:113-117

[11] Druzhinina IS, Seidl-Seiboth V, Herrera-Estrella A, Horwitz BA, Kenerley CM, Monte E, et al. *Trichoderma*: The genomics of opportunistic success. Nature Reviews. Microbiology. 2011;**9**:749-759

[12] Mukherjee PK, Buensanteai N, Morán-Diez ME, Druzhinina I, Kenerley CM. Functional analysis of non-ribosomal peptide synthetases (NRPSs) in *Trichoderma virens* reveals a polyketide synthase (PKS)/NRPS hybrid enzyme involved in the induced systemic resistance response in maize. Microbiology. 2012;**158**:155-165

[13] Harman GE. *Trichoderma*-not just for biocontrol anymore. Phytoparasitica. 2011;**39**:103-108

[14] Cheverri P, Gazis RO, Samuels GJ. *Trichoderma amazonicum*, a new endophytic species on *Hevea brasiliensis* and *H. guianensis* from the Amazon basin. Mycologia. 2011;**103**:139-151

[15] Atonasova L, Crom S, Gruber S, Coulpier F, Seidl-Seiboth V, Kubicek C, et al. Comparative transcriptomics reveals different strategies of *Trichoderma* mycoparasitism. BMC Genomics. 2013;**14**:121

[16] Romao-Dumaresq AS, Luis de Araújo W, Talbot NJ, Thornton CR. RNA interference of endochitinases in the sugarcane endophyte *Trichoderma virens* 223 reduces its fitness as biocontrol

agent of pineapple disease. PLoS One. 2012;7:47888

[17] Lorito M, Woo SL, Harman GE, Monte E. Translational research on Trichoderma: From 'omics' to the field. Annual Review of Phytopathology. 2010;**48**:395-417

[18] Benítez T, Rincón A, Carmen Limón M. Biocontrol mechanisms of *Trichoderma* strains. International Microbiology. 2004;7:249-260

[19] Kubicek CP, Herrera-Estrella A, Seidl-Seiboth V, Martinez DA, Druzhinina IS. Comparative genome sequence analysis underscores mycoparasitism as the ancestral life style of *Trichoderma*. Genome Biology. 2011;**12**:R40

[20] Montero-Barrientos M, Hermosa R, Cardoza RE, Gutiérrez S, Monte E. Functional analysis of the *Trichoderma harzianum* nox1 gene, encoding an NADPH oxidase, relates production of reactive oxygen species to specific biocontrol activity against *Pythium ultimum*. Applied and Environmental Microbiology. 2011;**77**:3009-3016

[21] Mukherjee PK, Wiest A, Ruiz N, Keightley A, Moran-Diez ME, McCluskey K, et al. Two classes of new Peptaibols are synthesized by a single non-ribosomal peptide synthetase of *Trichoderma virens*. Journal of Biological Chemistry. 2011;**286**(6):544-554

[22] Gruber S, Kubicek CP, Seidl-Seiboth V. Differential regulation of orthologous chitinase genes in mycoparasitic *Trichoderma* species. Applied and Environmental Microbiology. 2011;**77**(20):7217-7226

[23] McIntyre M, Nielsen J, Arnau J. Proceedings of the 7th European Conference on Fungal Genetics. Denmark: Copenhagen; 2004

[24] Zamioudis C, Pieterse CM. Modulation of host immunity by beneficial microbes. Molecular Plant-Microbe Interactions. 2012;**25**:139-150

[25] Shoresh M, Harman GE, Mastouri F. Induced systemic resistance and plant responses to fungal biocontrol agents. Annual Review of Phytopathology. 2010;**48**:21-43

[26] Azarmi R, Hajieghrari B, Giglou A. Effect of *Trichoderma* isolates on tomato seedling growth response and nutrient uptake. African Journal of Biotechnology. 2011;**10**:5850-5855

[27] Doni F, Anizan I, Che Radziah CMZ, Salman AH, Rodzihan MH, Wan Mohtar WY. Enhancement of rice seed germination and vigour by *Trichoderma spp*. Research Journal of Applied Sciences, Engineering and Technology. 2014;**7**(21):4547-4552

[28] Tjamos EC, Tjamos SE, Asntoniou PP. Biological management of plant diseases: Highlights on research and application of biological management. Journal of Plant Pathology. 2010;**92**:17-22

[29] Vinale F, Sivasithamparam K, Ghisalberti EL. Trichoderma secondary metabolites that affect plant metabolism. Natural Product Communications. 2012;**7**:1545-1550

[30] Vinale F, Sivasithamparam K, Ghisalberti EL, Woo S, Nigro ML, Marra R, et al. *Trichoderma* secondary metabolites active on plant and fungal pathogens. The Open Mycology Journal. 2014;**8**:127-139

[31] Salas-Marina MA, Silva-Flores MA, Uresti-Rivera EE, Castro-Longoria E, Herrera-Estrella A, Casas-Flores S. Colonization of Arabidopsis roots by *Trichoderma atroviride* promotes growth and enhances systemic disease resistance through jasmonic acid ethylene and salicylic acid pathways.

European Journal of Plant Pathology. 2011;**131**:15-26

[32] Gravel V, Antoun V, Tweddell RJ. Growth stimulation and fruit yield improvement of greenhouse tomato plants by inoculation with *Pseudomonas putida* or *Trichoderma atroviride*: Possible role of indoleacetic acid (IAA). Soil Biology and Biochemistry. 2007;**39**:1968-1977

[33] Viterbo A, Horwitz BA. Mycoparasitism. In: Borkovich K, Ebbole DJ, editors. Cellular and Molecular Biology of Filamentous Fungi. Washington, DC: ASM Press; 2010. pp. 676-693

[34] Contreras-Cornejo HA, Macías-Rodríguez L, Beltrán-Peña E, Herrera-Estrella A, López-Bucio J. *Trichoderma* induced plant immunity likely involves both hormonal- and camalexin dependent mechanisms in *Arabidopsis thaliana* and confers resistance against necrotrophic fungus *Botrytis cinerea*. Plant Signaling & Behavior. 2011;**6**:1554-1563

[35] Martínez-medina A, Del M, Alguacil M. Phytohormone profiles induced by trichoderma isolates correspond with their biocontrol and plant growth-promoting activity on melon plants. Journal of Chemical Ecology. 2014;**40**:804-815

[36] Tucci M, Ruocco M, De Masi L, De Palma M, Lorito M. The beneficial effect of *Trichoderma* spp. on tomato is modulated by the plant genotype. Molecular Plant Pathology. 2010;**12**:341-354

[37] Hoyos-Carvajal L, Bissett J. Biodiversity of *Trichoderma* in neotropics. In: Grillo O, Venora G, editors. The Dynamical Processes of Biodiversity—Case Studies of Evolution and Spatial Distribution. Rijeka: Intech; 2011. pp. 303-320

[38] Saravanakumar K, Arasu VS, Kathiresan K. Effect of *Trichoderma* on soil phosphate solubilization and growth improvement of *Avicennia marina*. Aquatic Botany. 2012;**104**:101-105

[39] Umashankar N, Venkateshamurthy P, Krishnamurthy R, Raveendra HR, Satish KM. Effect of microbial inoculants on the growth of silver oak (*Grevillea robusta*) in nursery condition. International Journal of Environmental Science and Development. 2012;**3**(1):72-76

[40] Tchameni SN, Ngonkeu MEL, Begoude BAD, Nana LW, Fokom R, Owona AD. Effect of *Trichoderma asperellum* and arbuscular mycorrhizal fungi on cacao growth and resistance against black pod disease. Crop Protection. 2011;**30**(10):1321-1327

[41] Coudert Y, Perin C, Coutois B, Khong NG, Gantet P. Genetic control of root development in rice, the model cereal. Trends in Plant Science. 2010;**15**:219-226

[42] Joshi BB, Bhatt RP, Bahukhandi D. Antagonistic and plant growth activity of *Trichoderma* isolates of Western Himalayas. Journal of Environmental Biology. 2010;**31**(6):921-928

[43] Mastouri F, Björkman T, Harman GE. Seed treatment with *Trichoderma harzianum* alleviates biotic, abiotic, and physiological stresses in germinating seeds and seedlings. Phytopathology. 2010;**100**(11):1213-1221

[44] Yang CA, Cheng CH, Lo CT, Liu SY, Lee JW, Peng KH. A novel L-amino acid oxidase from *Trichoderma harzianum* EST 323 associated with antagonism of *R. solani*. Journal of Agricultural and Food Chemistry. 2011;**59**:4519-4526

[45] Harman GE, Howell CR, Viterbo A, Chet I, Lorito M. *Trichoderma* plant symbionts species—Opportunistic,

avirulent plant symbionts. Nature Reviews. Microbiology. 2004;**2**:43-56

[46] Kachroo A, Robin GP. Systemic signaling during plant defense. Current Opinion in Plant Biology. 2013;**16**:527

[47] Sáenz-Mata J, Jiménez-Bremont JF. HR4 gene is induced in the Arabidopsis–*Trichoderma atroviride* beneficial interaction. International Journal of Molecular Sciences. 2012;**13**:9110-9128

[48] Schuster A, Schmoll M. Biology and biotechnology of Trichoderma. Applied Microbiology and Biotechnology. 2010;**87**(3):787-799

[49] Vinale F, Sivassithamparam K, Ghisalbert EL, Marra R, Woo SL, Larito M. *Trichoderma*-plant-pathogen interactions. Soil Biology & Biochemistry. 2008;**40**:1-10

[50] Viterbo A, Wiest A, Brotman Y, Chet I, Kenerley C. The 18mer peptaibols from *Trichoderma virens* elicit plant defense responses. Molecular Plant Pathology. 2007;**8**:737-746

[51] Ruocco M, Lanzuise S, Vinale F, Marra R, Turrà T, Woo SL, et al. Identification of a new biocontrol gene in *Trichoderma atroviride*: The role of an ABC transporter membrane pump in the interaction with different plant-pathogenic fungi. Molecular Plant-Microbe Interactions. 2009;**22**:291-301

[52] Gajera H, Domadiya R, Patel S, Kapopara M, Golakiya B. Molecular mechanism of *Trichoderma* as bio-control agents against phytopathogen system—A review. Current Research in Microbiology and Biotechnology. 2013;**1**(4):133-142

[53] Osbourn A. Secondary metabolic gene clusters: Evolutionary toolkits for chemical innovation. Trends in Genetics. 2010;**26**:449-457

[54] Mukherjee PK, Horwitz BA, Kenerley CM. Secondary metabolism in *Trichoderma* a genomic perspective. Microbiology. 2012b;**158**:35-45

[55] Baker SE, Perrone G, Richardson NM, Gallo A, Kubicek CP. Phylogenetic analysis of polyketide synthase-encoding genes in *Trichoderma*. Microbiology. 2012;**158**:35-45

[56] Shi M, Chen L, Wang XW, Zhang T, Zhao PB, Song XY, et al. Antimicrobial peptaibols from *Trichoderma pseudokoningii* induce programmed cell death in plant fungal pathogens. Microbiology. 2012;**158**:166-175

[57] Keswani C, Mishra S, Sarma B. Unraveling the efficient application of secondary metabolites of various *Trichoderma spp*. Applied Microbiology and Biotechnology. 2014;**98**:533-544

[58] Jacobsen BJ. Biological control of plant disease by phyllosphere applied biological control agents. In: Baily M, editor. Phyll. Microbiology. London: CRC Press; 2006. pp. 135-149

[59] Reino JL, Guerrero RF, Hernandez-Galan R, Collado IG. Secondary metabolites from species of the biocontrol agent *Trichoderma*. Phytochemistry. 2008;**7**:89-123

[60] Degenkolb T, Dieckmann R, Nielsen KF, Gräfenhan T, Theis C, Zafari H, et al. The *Trichoderma brevicompatum* clade: A separate lineage with new species, new peptaibiotics, and mycotoxins. Mycological Progress. 2008;**7**:177-219

[61] Szekeres A, Kredics L, Antal Z, et al. Isolation and characterization of protease overproducing mutants of *Trichoderma harzianum*. FEMS Microbiology Letters. 2004;**233**:215-222

[62] Degenkolb T, Karimi AR, Dieckmann R, Neuhof T, Baker SE. The production of multiple small

peptaibol families by single 14-module peptide synthetases in *Trichoderma/Hypocrea*. Chemistry & Biodiversity. 2012;**9**:499-535

[63] Mora'n-Diez E, Hermosa R, Ambrosino P, Cardoza RE, Gutie'rrez S, Lorito M, et al. The ThPG1 endopolygalacturonase is required for the *Trichoderma harzianum*–plant beneficial interaction. Molecular Plant Microbe Interaction. 2009;**22**:1021-1031

[64] Vos CMF, De Cremer K, Cammue BPA, De Coninck B. The toolbox of *Trichoderma* spp. in biocontrol of Botrytis cinerea disease. Molecular Plant Pathology. 2015;**16**:400-412

[65] Singh R, Ong-Abdullah M, Low ET, Manaf MA, Rosli R, Nookiah R, et al. Oil palm genome sequence reveals divergence of interfertile species in old and new worlds. Nature. 2013;**500**:335-339

[66] Chaube HS, Pundhir VS. Crop Diseases and their Management. PHI Learning Private Limited: New Delhi; 2012

[67] Segarra G, Casanova E, Aviles M, Trillas I. *Trichoderma asperellum* T34 controls Fusarium with disease in tomato plant in soilless culture through competition for iron. Microbial Ecology. 2010;**59**:141-149

[68] Miethke M. Molecular strategies of microbial iron assimilation from high-affinity complexes to cofactor assembly systems. Metallomics. 2013;**5**:15-28

[69] Lehner SM, Atanasova L, Neumann NK, Krska R, Lemmens M. Isotope-assisted screening for iron-containing metabolites reveals high diversity among known and unknown siderophores produced by *Trichoderma* spp. Applied and Environmental Microbiology. 2013;**79**:18-31

[70] Nassr S, Barakat R. Effect of factors on conidium germination of *Botrytis cinerea* in vitro. International Journal of Plant and Soil Science. 2013;**2**(1):41-45

[71] Bae H, Roberts DP, Lim H-S, Strem MD, Park SC, Ryu CM, et al. Endophytic *Trichoderma* isolates from tropical environments delay disease onset and induce resistance against *Phytophthora capsici* in hot pepper using multiple mechanisms. Molecular Plant-Microbe Interactions. 2011;**24**:336-351

Chapter 6

Trichoderma harzianum Rifai: A Beneficial Fungus for Growth and Development of *Abroma augusta* L. Seedlings with Other Microbial Bio-Inoculants

Vipin Parkash, Akshita Gaur, Rahul Agnihotri and Ashok Aggarwal

Abstract

Rhizospheric microbes play an important role in plant health. Rhizosphere is an area around the roots of plants where all microbes repose and influence the health of plants. These microbes require organic matter for their activity and provide nutrients to the plants and maintain the plant health. In this research paper, these useful microbes like fungi (*Trichoderma harzianum*), endomycorrhizae (Arbuscular Mycorrhiza) and bacteria (*Pseudomonas putida*) were isolated and after mass multiplication applied as bio-inoculants in alone and in combination to see the effect on growth and development of *Abroma augusta* seedlings which is a threatened medicinal plant in north-eastern part of India. *T. harzianum* alone and in combined form showed significant growth and development effect on seedlings. The effect of alone and combined treatments of *T. harzianum* on growth and development of this important medicinal plant species has been discussed in detail in this research paper.

Keywords: endomycorrhizae, microbes, *Pseudomonas putida*, rhizosphere, threatened medicinal plant

1. Introduction

Abroma augusta L. or Devil's Cotton is commonly known as *Ulat Kambal* (*Pisachkarpas* in Ayurvedic system, *Gorokhia koroi* in Assam, *Dadhubedang* in Meghalaya) when translated into English it means the inner side of a blanket. This is possibly because the bark of the tree yields a beautiful silky fiber, like that of hemp. Root bark of *Ulat Kambal* is a valuable emmenagogue and uterine tonic, chiefly used in intra-uterine diseases and other gynecological disorders mostly related to menstrual disorders such as dysmenorrhoea, amenorrhoea and gonorrhea. Powdered root of this medicinal plant is an abortifacient and anti-fertility agent. The leaves and stem are demulcent. Infusion of fresh leaves and stems is effective in treatment of gonorrhea [1].

IntechOpen

Abroma augusta L. is an evergreen small tree and native of Asia. In India, it is found in eastern and western Himalayan region. Plant reaches 10 feet (2.5 m) in height with very little spread. The leaves are heterophyllic, reach 8 inches (20 cm) across and are 3–5 lobed with very distinct palmate veins. The leaves and stems are covered with soft bristly hairs that are very irritating to the touch. The bark yields a jute-like fiber. The plants bloom from late spring to early summer. Dark maroon flowers are formed in terminal panicles. Individual flowers are up to 3 inches (7.5 cm) across. *A. augusta* is also featured as "Plant of the Week" May 28–June 4, 2004 [2]. This plant species is present in Assam, Meghalaya, Arunachal Pradesh and Nagaland states of north eastern Himalayas. The plant species is declining in number due to exploitation and lack of awareness and also facing severe threat due to biotic pressure [3]. According to one of report [4], there are only 2000 medicinal plants known in Brahmaputra valley and Assam hills in North East India in which important ones are *A. augusta*, *Smilax glabra, Aquilaria malaccensis* and *Hydnocarpus kurzii*. The root and stem bark of this plant is in great demand in herbal market (800–1000/kg) as per a report by Singh [5]. This plant also needs protection according to one of report by Botanical Survey of India at Itanagar, Arunachal Pradesh [4]. 'Monofil', an herbal formulation containing *A. augusta* can be used as an alternate in the treatment of post menopause syndrome [6]. Also, *A. augusta* harbors a good population of microbes in its rhizosphere.

A narrow zone of soil affected by the presence of plant roots is defined as rhizosphere [7]. The rhizosphere is an environment that the plant itself helps to create and where pathogenic and beneficial microorganisms constitute a major influential force on plant growth and health [8]. Microbial groups and other bio-agents found in the rhizosphere include bacteria, fungi, nematodes, protozoa, algae and microarthropods [8, 9]. Microorganisms that adversely affect plant growth and health are the pathogenic fungi, oomycetes, bacteria and nematodes, whereas microorganisms that are beneficial include nitrogen-fixing bacteria, endo- and ectomycorrhizal fungi, plant growth-promoting rhizobacteria (PGPR) and fungi [10]. *Trichoderma* is the most common and easily culturable soil fungus present in almost all types of soil. *Trichoderma* has been reported to possess biocontrol activities against different pathogens and also possess plant growth promoting characteristics [11]. Similarly, some bacterial species are also known to help in plant growth promotion. One such Plant growth promoting rhizobacteria is *Pseudomonas putida* [12]. It is a gram negative bacterium which is widespread in nature and can be easily cultured. Arbuscular mycorrhizal (AM) fungi are known to increase phosphate solubilization and help increase in nutrient uptake thus, helping in improving overall growth of a plant. Also, interaction of AM fungi with *Trichoderma* sp. has been reported to have better effects in improving the overall growth of *Eucalyptus saligna* Sm. [13].

So keeping in mind all these biological and chemical processes, the study was undertaken on biodiversity of rhizospheric microbes and their effect on the seedlings of *Abroma augusta* which is a threatened important plant species in Brahmaputra Valley of Assam, India.

2. Materials and methods

2.1 Survey and collection of soil samples

As *Abroma augusta* L. is scattered in its distribution the survey of different study sites (Assam) were done for the collection of rhizospheric soil samples. The plant specimen was preserved in herbarium sheets for identification. Rhizosphere soil samples (at least three samples) were taken by digging out a small amount of soil (500 g) close to plant roots up to the depth of 15–30 cm and these samples were kept in sterilized polythene bags at 10°C for further processing in the laboratory.

2.2 Isolation of useful microbes

For qualitative studies of soil mycoflora, Warcup soil plate method [14] and Waksman soil dilution method [15] were used.

2.3 Isolation of *Trichoderma* inoculum

Fungal species *Trichoderma harzianum* was isolated from the soil samples by using serial soil dilution method [15] on potato dextrose agar (PDA) medium. The inoculated plates were incubated at 30°C for 4 days. The pure fungal colonies were picked up and purified by streaking on agar slants and incubated at 30°C for 7–8 days. Green conidia forming fungal bodies were selected and microscopic observation was done and the fungus was identified to be *T. harzianum* (Isolate/accession no. MSML/RFRI/TH-13). The preserved fungal isolate/culture maintained on PDA slants are retained with Mycology and Soil microbiology Laboratory, Rain Forest Research Institute, Jorhat, Assam, India for further study.

2.4 Preparation of solid substrate media

In this experiment, different saw dusts like *Pinus kesiya* Royle ex. Gordon, *Shorea robusta* Gaertn. and *Callicarpa arborea* Roxb. were taken for evaluation. The different saw dusts were shade dried. The dried saw dusts were mixed with wheat bran by adding sterilized water in the ratio (wheat bran: saw-dust: water; 3:1:4 w/w) as explained above. The moisture level of the mixture was maintained up to 50–60%. The substrate was sterilized through Autoclave (Labotech, BDI-81 make, India) at 120°C and 15lbs [16].

2.5 Mass multiplication of fungal inoculum

The inoculum of *Trichoderma harzianum* was grown on synthetic PDA (Potato Dextrose Agar) medium (SRL, India) for 7–8 days and incubated at 27–30° ± 1°C. The inoculum was kept in B.O.D. incubator (Labotech, BDI-55 make, India) for 10–12 days for maximum growth and sporulation. Then the inoculum containing medium was cut into small discs and were put in flasks containing wheat bran and different saw-dust medium in the ratio (3,1,4 w/w) for mass production of *T. harzianum*. Approximate 50 g of substrate was taken in 500 ml conical flasks, inoculated with 5 mm mycelial mat incubated at 28°C incubator for 7–10 days earlier. In control set, no saw dust component was added to the substrate. Six replicates of each treatment were taken. The colony forming units were calculated with help of the following formula through serial dilution of 1 g of substrate and results are expressed as cfu g^{-1} ml^{-1} of suspension of each substrate [16].

$$\mathrm{CFU/g/ml} = \frac{\text{Number of colonies per ml plated}}{\text{Total dilution factor}} \tag{1}$$

2.6 Mass multiplication of bacterial inoculum

The *Pseudomonas putida* (MSML/RFRI/Ps-1) multiplication was carried out through bacterial cultivation technique by using growth curve after specific time intervals of 1 h [17]. The inoculum of Bacteria was taken in stationary phase (10–11 h) for inoculation experiment.

2.7 Isolation of VAM = AM spores

Isolation of VAM spores was done by using wet sieving and decanting technique of Gerdemann and Nicolson [18] and Singh and Tiwari [19]. Sieves of different sizes, i.e. 150, 120, 90, 63 and 45 μm were used. About 150 μm sieve were used for collecting plant debris in the soil. About 100 g of soil were mixed in water in a small plastic container having the capacity of about 1500 ml. The soil was thoroughly mixed with water and allowed to settle for overnight. The water was decanted on a series of sieves in the following order 150, 120, 90, 63 and 45 μm from top to bottom on which spores were trapped. The trapped spores were then transferred to Whatman filter paper No. 1 by repeated washing with water. The spores were picked by hypodermic needle under stereo-binocular microscope and mounted in polyvinyl lactic acid.

2.8 Mass multiplication of endomycorrhizal inoculum

The mycorrhizal inoculum production was done by using '*soil funnel technique*'. Single dominant and efficient AM spore/s of *Glomus* sp. (MSML/RFRI/M_1) and (*Acaulospora* sp. (MSML/RFRI/M_2) was/were mass produced in this technique.

The best host were selected for starter culture of inoculum production, i.e. sorghum (*Sorghum vulgare*), wheat (*Triticum aestivum*) and onion (*Allium sativum*). In this technique, glass funnels/earthen funnels were taken and germination of seeds was made in such a way that roots of the seedlings must touch the inoculum of AM fungi. The seedlings were raised up to 30 days in the glass/earthen funnels containing sterilized sand: soil (40:120 g.). The experiment was repeated again and AM spores were collected by wet sieving and decanting technique of Gerdemann and Nicolson [18] and Singh and Tiwari [19] after 45–60 days. These final spores were used for mass multiplication by using different hosts in bigger earthen pots for further study.

To have mass inoculum in bulk quantity, the test inocula were multiplied in field conditions by preparing standard size beds on thin polyethylene sheet (0.5 mm) so that no contamination will occur to the inocula. The experiment was repeated for maintaining the inocula cultures viable for further experiments.

2.9 Cultivation, conservation and growth studies

The plantlets of target plant species after survey were raised from seeds/propagules with the help of inoculation in root trainers in laboratory conditions. These inoculated seedlings were transplanted in bigger pots and then in field conditions again with bioagents inoculation for primary establishment and better growth of quality seedlings. The experiment was designed in different treatments like single and double inoculations. Three replications of each treatment were taken. In control sets, no bioagent (inoculum) was given or added. The following design of experiment was adopted.

T_0 = Control (no inoculation).

T_1 = Treatment/inoculation of A endomycorrhizal strain (AM fungus).

T_2 = Treatment/inoculation of B another bio-agent (fungus).

T_3 = Treatment/inoculation of C another bio-agent (bacteria).

T_4 = Treatment/inoculation of A+B, A+C, B+C and A+B+C (endomycorrhizal strain + another bio-agent/s) (bioagents consortium).

Observations were recorded to see the inoculation effect on plant seedlings for the following parameters after specific time intervals up to 180 days after inoculation/DAI in sterilized soil conditions.

- Shoot length in cm (increase in height).
- Diameter in mm.
- Percentage root colonization.
- Mycorrhizal (AM) spore number.

3. Results

The data on the effect of bio-agents inoculation on height of *Abroma augusta* after inoculation (DAI)* in sterilized soil condition was tabulated and further analyzed (**Table 1**). The data on the increase in height (cm) was recorded after 30, 60,

Treatments	Initial height (cm)	Increase in height (cm)				
		30 days	60 days	90 days	120 days	150 days
Control	3.67 ± 0.76	5 ± 0.54	14.66 ± 2.0	16.33 ± 1.7	18.16 ± 0.95	19.33 ± 0.71
T_M	6.16 ± 1.06	2.84 ± 0.94	19.64 ± 1.3	21.51 ± 1.44	23.17 ± 1.34	24.17 ± 1.65
T_B	5.67 ± 0.14	4.16 ± 0.14	22.33 ± 5.3	23.33 ± 1.89	25.33 ± 1.93	26.33 ± 1.69
T_F	3 ± 0.41	5.5 ± 0.85	29.67 ± 0.54	31 ± 0.47	34 ± 0.47	35.33 ± 0.72
T_{M+B}	8.5 ± 0.62	3 ± 1.55	14 ± 4.55	15.17 ± 4.75	16.27 ± 4.75	18 ± 4.63
T_{M+F}	6.17 ± 1.6	4.83 ± 0.82	17.83 ± 1.25	18.5 ± 1.44	20.83 ± 1.78	21.83 ± 1.7
T_{F+B}	2.3 ± 0.29	4.53 ± 1.34	11.2 ± 4.94	12.87 ± 5.14	14.2 ± 5.4	15.27 ± 5.43
T_{M+B+F}	3.17 ± 0.14	5.76 ± 1.08	19.5 ± 2.33	21.5 ± 2.43	23.16 ± 1.33	23.83 ± 2.62
CV (%)	0.13	0.22	0.17	0.15	0.13	0.12

*Average of three replications (Days after inoculation).
M, mycorrhiza (consortium); B, bacteria (*Pseudomonas putida*); F, fungi (*Trichoderma harzianum*); CV, coefficient of variance; ±SEm, standard error of mean.

Table 1.
Effect of bio-agents inoculation on height of Abroma augusta *L. after inoculation (DAI)* in sterilized soil condition.*

Treatments	Initial diameter (mm)	Increase in diameter (mm)				
		30 days	60 days	90 days	120 days	150 days
Control	1.2 ± 0.094	0.63 ± 0.136	1.2 ± 0.17	1.57 ± 0.12	1.87 ± 0.24	1.97 ± 0.31
T_M	0.8 ± 0.047	0.33 ± 0.072	1.37 ± 0.22	2.13 ± 0.19	2.53 ± 0.17	2.66 ± 0.17
T_B	0.9 ± 0.045	1.1 ± 0	1.27 ± 0.25	2.3 ± 0.08	2.67 ± 0.03	2.93 ± 0.07
T_F	1.03 ± 0.072	0.5 ± 0.119	1.8 ± 0.14	2.27 ± 0.14	2.47 ± 0	2.6 ± 0.18
T_{M+B}	1.27 ± 0.098	0.3 ± 0.191	0.46 ± 0.3	1.16 ± 0.18	1.66 ± 0.19	1.79 ± 0.29
T_{M+F}	0.90 ± 0.043	0.93 ± 0.071	2.17 ± 0.19	2.77 ± 0.07	3.2 ± 0.08	3.4 ± 0.12

Treatments	Initial diameter (mm)	Increase in diameter (mm)				
		30 days	60 days	90 days	120 days	150 days
T_{F+B}	0.87 ± 0.075	0.6 ± 0.196	1.13 ± 0.43	1.46 ± 0.59	1.83 ± 0.67	2.29 ± 0.64
T_{M+B+F}	0.90 ± 0.091	0.87 ± 0.129	1.97 ± 0.11	3.0 ± 0.08	3.2 ± 0.14	3.43 ± 0.14
CV (%)	0.07	0.23	0.22	0.11	0.09	0.10

Average of three replications (Days after inoculation).
M, mycorrhiza (consortium); B, bacteria *(Pseudomonas putida)*; F, fungi *(Trichoderma harzianum)*; CV, coefficient of variance; ±SEm, standard error of mean.

Table 2.
Effect of bio-agents inoculation on diameter of Abroma augusta *L. after inoculation (DAI)* in sterilized soil condition.*

Treatments	Abiotic parameters		Biotic variables				
	Temperature (°C)	*pH*	Increase in height (cm)	Spore count/ 50 g	Hyphal infection (%)	Arbuscular infection (%)	Vesicular infection (%)
Control	34 ± 0	5.62 ± 0.25	29 ± 0.82	19 ± 0.30	35 ± 1.24	0	10 ± 0.37
T_M	34 ± 0	5.4 ± 0.47	35.33 ± 0.29	81 ± 0.072	100 ± 0	40 ± 0.19	50 ± 1.24
T_B	34 ± 0	5.43 ± 0.37	35.67 ± 0.16	4 ± 0.16	0 ± 0	0 ± 0	0 ± 0
T_F	34 ± 0	5.47 ± 0.19	41.33 ± 0.22	30 ± 0.44	30 ± 1.24	10 ± 0.072	20 ± 1.63
T_{M+B}	34 ± 0	5.62 ± 1.63	30.67 ± 1.63	28 ± 0.25	80 ± 0.22	40 ± 0.10	10 ± 0.29
T_{M+F}	34 ± 0	5.65 ± 0.22	37.67 ± 0.30	34 ± 1.24	70 ± 0.44	30 ± 0.25	20 ± 0.47
T_{F+B}	34 ± 0	5.6 ± 0.16	24.33 ± 0.072	10 ± 0.10	20 ± 0.29	0 ± 0	0 ± 0
T_{M+B+F}	34 ± 0	5.55 ± 0.82	29.67 ± 1.24	117 ± 0.82	100 ± 0	30 ± 0.37	60 ± 0.19
CV (%)	0	0.092	0.019	0.010	0.008	0.013	0.026

Average of three replications (Days after inoculation).
M, mycorrhiza (consortium); B, bacteria *(Pseudomonas putida)*; F, fungi *(Trichoderma harzianum)*; CV, coefficient of variance; ±SEm, standard error of mean.

Table 3.
Effect of bio-agents inoculation on Abroma augusta *L. after 180 days of inoculation (DAI)* in sterilized soil condition.*

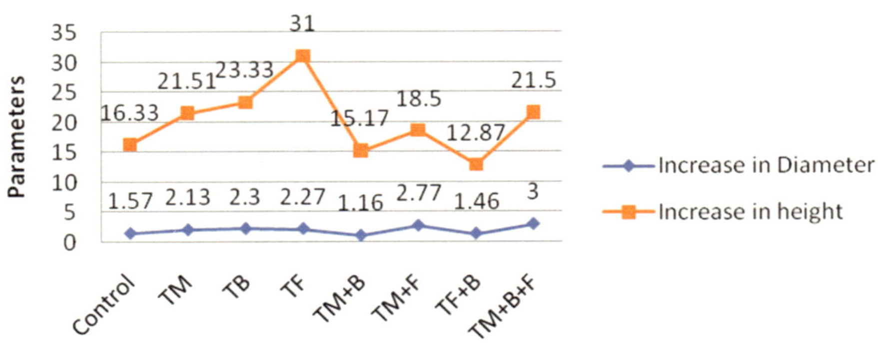

Figure 1.
Effect of bio-agents inoculation on increase in diameter and height of Abroma augusta *L. after 90 days (DAI)*.*

90, 120 and 150 days after inoculation. The increase in height after 30 days showed maximum height increase (5.76 ± 1.08) in $T_{M+B+F,}$ whereas, minimum increase in height (2.84 ± 0.94) was recorded in T_M treatment. The coefficient of variance was found to be 0.22 after analysis. After 60 days of inoculation, the increase in height (29.67 ± 0.54) was maximum in $T_{F,}$ treatment, whereas, minimum increase in height (11.2 ± 4.94) was recorded in T_{F+B} treatment. The coefficient of variance was 0.17. After 90 days of inoculation, $T_{F,}$ treatment showed maximum increase in

Figure 2.
(A–C) Trap and starter cultures of AM fungi inoculum. (D–G) Roots showing trapping of AM spores and harvesting of inoculums. (H–J) Mass production of Trichoderma harzianum *inoculums. (K–N) Mass production of* Pseudomonas putida *inoculums. (O–Q) Bioinoculation growth effect on seedlings of* Abroma augusta.

height (31 ± 0.47), whereas, minimum increase in height (12.87 ± 5.14) was found in T_{F+B} treatment. The coefficient of variance was found 0.15 in this case. The data after 120 days of inoculation revealed that $T_{F,}$ showed maximum increase in height (34 ± 0.47), whereas, T_{F+B} showed minimum increase in height (14.2 ± 5.4). The coefficient of variance was 0.13. After 150 days of inoculation, maximum increase in height (35.33 ± 0.72) was recorded in T_F treatment, whereas, minimum increase in height (15.27 ± 5.43) was recorded in $T_{F+B,}$ treatment. The coefficient of variance was 0.12 in this analysis (see **Table 1**).

The study of the effect of bio-agents inoculation on diameter of *Abroma augusta* after inoculation (DAI)* in sterilized soil condition in **Table 2**. The data on the increase in diameter (cm) was recorded after 30, 60, 90, 120 and 150 days after inoculation. The increase in diameter after 30 days showed maximum increase of (1.1 ± 0) in T_B treatment, whereas, minimum increase in diameter (0.3 ± 0.191) was recorded in T_{M+B} treatment. The coefficient of variance was found 0.23. After 60 days of inoculation, the increase in diameter was maximum (2.17 ± 0.19) in T_{M+F} treatment, whereas, minimum increase in diameter (0.46 ± 0.3) was recorded in T_{M+B} treatment. The coefficient of variance was 0.22. After 90 days of inoculation, $T_{M+B+F,}$ treatment showed maximum increase in diameter (3.0 ± 0.08), whereas, minimum increase in diameter (1.16 ± 0.18) was found in $T_{M+B,}$ treatment. The coefficient of variance was 0.11. The data after 120 days of inoculation revealed that, $T_{M+B+F,}$ showed maximum increase in diameter (3.2 ± 0.14), whereas, T_{M+B} treatment showed minimum increase in diameter (1.66 ± 0.19). The coefficient of variance was 0.09. After 150 days of inoculation, maximum increase in diameter (3.43 ± 0.14) was recorded in T_{M+B+F} treatment, whereas, minimum increase in diameter (1.79 ± 0.29) was recorded in $T_{M+B,}$ treatment. The coefficient of variance was 0.10 (see **Table 2**).

The data on the effect of bio-agents inoculation on *Abroma augusta* after 180 days of first stage of inoculation (DAI)* (before second stage inoculation) was tabulated in **Table 3**. The analysis of the data revealed that among the various treatments the pH varies between 5.4 ± 0.47 to 5.65 ± 0.22 respectively. The increase in height (cm) was recorded and it was found that, T_F treatment showed maximum increase in height (41.33 ± 0.22), whereas, minimum increase in height (24.33 ± 0.072) was recorded in T_{F+B} treatment (**Figures 1** and **2**).

4. Discussion

According to Jha et al. [20], as they have already reported that the ecology of microorganisms, however, cannot be considered solely in terms of their relationships with the abiotic environment, as their ability to co-exist with other microorganisms. However, the link between rhizospheric microbial biota and abiotic soil properties can be exclusively advocated for a broader utilization as an ecological parameter for any plant species. Therefore, the present study was an attempt to determine the association ecology of an important plant species like *Abroma augusta* under threat due to anthropogenic pressure in the Brahmaputra valley of Assam, India. The synergistic effect of dual inoculation of AM and *Rhizobium* on chickpea was found on nodulation, plant growth, dry matter production and N-fixation by Jalali and Thareja [21]. Similarly in cowpea, inoculation with AM and *Rhizobium* under field conditions increased shoot dry matter and yield over the alone AM or *Rhizobium* inoculation [22]. Similar synergistic effect of dual inoculation was reported in *Leucaena leucocephala* and *Cajanus cajan*. The combined inoculations of symbionts showed significant increased N-fixations growth and nutrient uptake in *Leucaena leucocephala* and *Cajanus cajan* [23, 24]. Similarly in this case, the combined synergistic effect of tripartite inoculation treatment showed positive growth effect on *Abroma augusta* seedlings.

According to Chang [25] and Rani et al. [26], the above ground part, i.e. height of plant and fresh shoot weight were more in *Trichoderma* sp. treatment in *Acacia nilotica*. In the present study also, above ground part, i.e. shoot length were more in *Trichoderma* treatment alone in *Abroma augusta* seedlings. It is may be due to *Trichoderma* fungus which is secreting some substances in the rhizosphere which are responsible for better growth of above ground parts.

In the present investigation, results were similar to the findings of Gill and Singh [27], Parkash and Aggarwal [28], Parkash et al. [29]. They reported that the mutualistic association was accounted for better colonization and plant growth due to interchange of carbon, phosphate and nitrogen between host fungi and bacteria. The dual inoculation of AM fungi and *Rhizobium* had synergistic effect on nodulation, plant growth, dry matter, production and nitrogen fixation [30, 31]. Similarly, Singh and Singh [32] also recorded higher yield in dual inoculation of VAM and *Rhizobium* in lentil. Increased N-fixation due to increased mycorrhizal colonization and nodulation may contribute towards growth and yield of plants [27]. Kaushish et al. [33] also found mycorrhizal inoculation effect on growth and physical parameters on *Rauwolfia serpentina* Benth. Ex. Kurtz. Similar results on physical parameters were found by synergistic inoculation on *Abroma augusta* [34].

Similar results were also obtained by Kumar et al. [35]. They reported that the inoculation with all the tested VAM fungi, e.g., *Glomus mosseae, G. fasciculatum, G. constrictum, Acaulospora laevis* and *Gigaspora gilmorei* resulted in significant increase in plant height of chickpea because in present investigation *G. mosseae* in combination showed better growth response in *Abroma augusta* after 90 days.

The root colonization by mycorrhiza was directly related to nutrient uptake by plants. Lowest nutrient uptake was observed in non-mycorrhizal plants and highest nutrient in mycorrhizal plants grown in sterilized soil. Numbers of mycorrhizal spores were also higher in mycorrhizal inoculated soil and directly related to mycorrhizal root colonization. The plants with highest root colonization showed greater number of mycorrhizal spores in soil [36]. In the present investigation, same trend was observed in the plant seedlings as far as mycorrhizal spore number and root colonization were concerned.

5. Conclusion

All the bio-inoculation treatments were significant than control treatment in terms of increase in height and girth. The combined treatments T_{F+B} and T_{M+B+F} are the best and significant bio-inoculation treatments for this plant species. If all parameters are considered together then consortium of endomycorrhizae, fungus, bacterium (T_{M+B+F}) treatment is better in inoculating the target species although, all alone bioagents had good growth and development effect on the seedlings than control treatment where no inoculum was added. Although, bacterium (*Pseudomonas putida*) alone had no significant effect but if it mixed with other bioagents like *Trichoderma harzianum* and AM fungi consortium (*Glomus* species + *Acaulospora* species) then this treatment has significant effect on this target plant species.

Acknowledgements

The first author thanks the Indian Council of Forestry Research & Education (ICFRE), Dehradun for financial help under the sanctioned project no. RFRI-42/2010-2011/SFM.

Conflict of interest

We confirm that there are no conflicts of interest.

Notes/thanks/other declarations

First Author offers his honest obligation and regards to The Director, Rain Forest Research Institute (RFRI), Jorhat, Assam for his kind permission to carry out this important project and for his valuable suggestion and encouragement throughout the work.

I heartily thank the supporting staff especially Mr. Ankur Jyoti Saikia (TA) for valuable help and assistance at various stages of experimental and analytical work. I am also indebted to Ms. Adrita Borkatoki (JRF) and Mr. Debajyoti Bora (FA) under the project for their help rendered during the field survey and data typing & analyses.

Authors details

Vipin Parkash[1*], Akshita Gaur[1], Rahul Agnihotri[1] and Ashok Aggarwal[2]

1 Forest Pathology Discipline, Forest Protection Division, Forest Research Institute (Indian Council Forestry Research and Education, Autonomous Council under Ministry of Environment, Forest and Climate Change, Government of India), Dehradun, Uttarakhand, India

2 Botany Department, Kurukshetra University, Kurukshetra, Haryana, India

*Address all correspondence to: bhardwajvpnpark@rediffmail.com

References

[1] Ahmad W, Khan T, Ahmad B, Iqbal Z, Ahmad M, Jan Q, et al. Chemical composition of *Abroma augusta* Linn. seed oil. Pakistan Journal of Biological Science. 2003;**6**(12):1033-1034

[2] Available from: http://www.plantoftheweek.org [Accessed: Dec 2012, Jan 2015]

[3] Parkash V, Saikia AJ. Habitational abiotic environmental factors alter arbuscular mycorrhizal composition, species richness and diversity index in *Abroma augusta* L. (Malvaceae) rhizosphere. Plant Pathology & Quarantine. 2015;**5**(2):98-120

[4] Srivastava RC, Choudhary RK. Floristic scenario of Itanagar wildlife sanctuary—A case study. Bulletin of Arunachal Forest Research. 2006;**22** (1&2):17-21

[5] Singh AB. Medicinal plant survey of Dumka, Hazaribagh and Gumla districts. Forest Resources Survey Division, Jharkhand, Ranchi. 2003:25-26

[6] Nanda UK. P319 clinical evaluation of non-hormonal drug 'Minofil'in the management of post menopausal syndrome. Maturitas. 1996;**27**:215

[7] Hrynkiewicz K, Baum C. The potential of rhizosphere microorganisms to promote the plant growth in disturbed soils. In: Environmental Protection Strategies for Sustainable Development. Dordrecht: Springer; 2012. pp. 35-64

[8] Lynch J. The Rhizosphere. Wiley, London, UK; 1990. p. 458

[9] Raaijmakers JM. Rhizosphere and rhizosphere competence. In: Maloy OC, Murray TD, editors. Encyclopedia of Plant Pathology. USA: Wiley; 2001. pp. 859-860

[10] Raaijmakers JM, Paulitz TC, Steinberg C, Alabouvette C, Moënne-Loccoz Y. The rhizosphere: A playground and battlefield for soilborne pathogens and beneficial microorganisms. Plant and Soil. 2009; **321**(1–2):341-361

[11] Vinale F, Sivasithamparam K, Ghisalberti EL, Marra R, Woo SL, Lorito M. *Trichoderma*-plant-pathogen interactions. Soil Biology and Biochemistry. 2008;**40**(1):1-10

[12] Meliani A, Bensoltane A, Benidire L, Oufdou K. Plant growth-promotion and IAA secretion with *Pseudomonas fluorescens* and *Pseudomonas putida*. Research & Reviews: Journal of Botanical Sciences. 2017;**6**(2):16-24

[13] Parkash V, Aggarwal A. Interaction of VAM fungi with *Trichoderma viride* and *Rhizobium* species on establishment and growth of *Eucalyptus saligna* Sm. International Scientific Research Journal. 2011;**3**:200-209

[14] Warcup JH. The soil-plate method for isolation of fungi from soil. Nature. 1950 Jul;**166**(4211):117

[15] Waksman SA. Principle of Soil Microbiology. Baltimore, Maryland: Williams and Wilkins Co.; 1927. 897 p

[16] Parkash V, Saikia AJ. Production and multiplication of native compost fungal activator by using different substrates and its influence on growth and development of *Capsicum chinensis* Jacq."Bhut Jolokia". Biotechnology Research International. 2015;**2015**:1-7

[17] Novick A. Growth of bacteria. Annual Reviews in Microbiology. 1955; **9**(1):97-110

[18] Gerdemann JW, Nicolson TH. Spores of mycorrhizal endogone species

extracted from soil by wet sieving and decanting. Transactions of the British Mycological society. 1963;**46**(2):235-244

[19] Singh SS, Tiwari SC. Modified wet sieving and decanting technique for enhanced recovery of spores of vesicular arbuscular mycorrhizal (VAM) fungi in forest soils. Mycorrhiza News. 2001;**12**: 12-13

[20] Jha DK, Sharma GD, Mishra RR. Ecology of soil microflora and mycorrhizal symbionts in degraded forests at two altitudes. Biology and Fertility of Soils. 1992;**12**(4):272-278

[21] Jalali BL, Thareja ML. Plant growth responses to vesicular-arbuscular mycorrhizal inoculation in soils incorporated with rock phosphate. Indian Phytopathology. 1985;**38**(2): 306-310

[22] Islam R, Ayanaba A. Effect of seed inoculation and preinfecting cowpea (*Vigna unguiculata*) with *Glomus mosseae* on growth and seed yield of the plants under field conditions. Plant and Soil. 1981;**61**(3):341-350

[23] Punj V, Gupta RP. Response of subabul to inoculation with vesicular arbuscular mycorrhiza, phosphorus and nitrogen fertilizers. Agricultural Research Journal Punjab Agricultural University. 1989;**26**:450-456

[24] Sekhon GK, Gupta RP, Pandher MS, Arora JK. Symbiotic effectiveness of Hup+ *Rhizobium*, VAM fungi and phosphorus levels in relation to nitrogen fixation and plant growth of *Cajanus cajan*. Folia Microbiologica. 1992;**37**(3):210

[25] Chang Y. Increased growth of plants in the presence of the biological control agent *Trichoderma harzianum*. Plant Disease. 1986;**70**:145-148

[26] Rani P, Aggarwal A, Mehrotra RS. Growth responses in Acacia nilotica inoculated with VAM fungi (*Glomus mosseae*), *Rhizobium* sp and *Trichoderma harzianum*. Indian Phytopathology. 1999;**52**(2):151-153

[27] Gill TS, Singh RS. Effect of *Glomus fasciculatum* and *Rhizobium* inoculation on VA mycorrhizal colonization and plant growth of chickpea. Journal of Mycology and Plant Pathology. 2002; **32**(2):162-167

[28] Parkash V, Aggarwal A. Diversity of endomycorrhizal fungi and their synergistic effect on the growth of *Acacia catechu* Willd. Journal of Forest Science. 2009;**55**:461-468

[29] Parkash V, Sharma S, Aggarwal A. Symbiotic and synergistic efficacy of endomycorrhizae with *Dendrocalamus strictus* L. Plant, Soil and Environment. 2011;**57**(10):447-452

[30] Moawad H, El-Din SB, Abdel-Aziz RA. Improvement of biological nitrogen fixation in Egyptian winter legumes through better management of *Rhizobium*. Plant and Soil. 1998;**204**(1): 95-106

[31] Parkash V, Aggarwal A, Sharma V. Rhizospheric effect of vesicular arbuscular mycorrhizal inoculation on biomass production of *Ruta graveolens* L.: A potential medicinal and aromatic herb. Journal of Plant Nutrition. 2011; **34**(9):1386-1396

[32] Singh OS, Singh RS. Effects of phosphorus and *Glomus fasciculatus* inoculation on nitrogen fixation, P uptake and yield of lentil (*Lens culinaris* medic) grown on an unsterilized sandy soil. Environmental and Experimental Botany. 1986;**26**(2):185-190

[33] Kaushish S, Kumar A, Aggarwal A. Influence of hosts and substrates on mass multiplication of *Glomus mosseae*. African Journal of Agricultural Research. 2011;**6**(13):2971-2977

[34] Parkash V, Saikia AJ, Bora D. Locational variability of

physicochemical properties of rhizospheric soil and mycobiota associated with *Abroma augusta* L. Advanced Biomedical Research. 2014; **5**(3):156-165

[35] Kumar R, Jalali BL, Chand H. Influence of vesicular arbuscular mycorrhizal fungi on growth and nutrient uptake in chickpea. Journal of Mycology and Plant Pathology. 2002; **32**(1):11-15

[36] Hazarika DK, Das KK, Dubey LN. Effect of vesicular arbuscular mycorrhizal fungi inoculation on growth and nutrient uptake of black gram. Journal of Mycology and Plant Pathology. 1999;**29**(2):201-204